Farmers' Fowls

by New South Wales Dept. of Agriculture

with an introduction by Jackson Chambers

This work contains material that was originally published in 1907.

This publication is within the Public Domain.

This edition is reprinted for educational purposes
and in accordance with all applicable Federal Laws.

IMPORTANT NOTE & DISCLAIMER

Self Reliance Books

Get more historic titles on animal and stock breeding, gardening and old
fashioned skills by visiting us at:

http://selfreliancebooks.blogspot.com/

~ introduction ~

Here at **Self-Reliance Books** we are dedicated to bringing you the best in *dusty-old-book-knowledge* – this time, an old book on the production comparisons of egg-laying Hens.

This special edition of ***'Farmers' Fowls'*** was produced by the *New South Wales Dept. of Agriculture*, and first published in 1907, making it well over on e-hundred years old.

The book features sections on *The Wyandotte, Poultry Shows, Justification for New Breeds, Orpingtons, Plymouth Rocks, Langshans, Dorkins and Houdans, The Mediterranean Breeds, Houses, Coops, Nests, Drinking-Vessels, etc.,* and more.

An essential old book for all Poultry Breeders, and anybody considering taking the plunge and entering the business. This book is a *must*-read, particularly for all those interested in the Poultry industry in the historical aspect.

~ Jackson Chambers

State of Jefferson, April 2018

PREFACE.

POULTRY authorities, and other writers in the Australian Press, are cease-lessly reminding us of the enormous quantities of eggs and poultry imported into England annually, and telling the breeders of this country the excellent markets awaiting the products of our poultry yards in the far away world's market, unmindful or oblivious of the fact that the Australian Commonwealth, with all its resources in limitless acres, cheaply produced foods, almost perpetual sunshine, and other conditions favourable for poultry breeding, does not produce enough eggs for its own needs ; while, coming to our own State, which has the above and other essentials calculated to make it the poultry-man's paradise, we are in the unenviable position of not only being unable to supply our own egg demands, but have to regretfully acknowledge annual increasing imports, which culminated last year in the enormous total of 1,452,207 dozens, or about one dozen for every man, woman, and child in the State.

There are numerous causes for this enormous shortage in our supplies, the chief being that we have not enough small farmers, orchardists, other such holders, and cottagers, they being the chief poultry growers the world over. Another reason is that we are a very prosperous community, and many of the class who are the principal producers in other countries are too well off here to bother about fowls ; a third reason for our increasing imports being that we consume more eggs in proportion to our population than any other State in the Commonwealth, and this consumption is increasing at a greater rate than is the supply.

However, despite our ability and willingness to purchase rather than produce, it is to be regretted that with every essential for profitable poultry breeding, we are obliged to largely buy, and the chief purport of this work is the hope that those for whom it is intended may, through its perusal, become better acquainted with the profitableness of the hen, and so improve and increase their stock to the end of at first reducing, and finally wiping out, the immense imports, which last year were valued at £37,000 ; and when this big task has been accomplished, there is awaiting us the insatiable English market, which, if not so remunerative as our own, has the important and paramount feature of permanency, any quantity that we may ever hope to send being only a drop in the great ocean of English imports—a sample amongst the bulk quantities which many countries and nationalities are profitably producing for the world's market.

G. BRADSHAW.

December, 1906.

INDEX.

LIST OF CONTENTS.

PLATES.

FROM AGRICULTURAL GAZETTE OF N. S. WALES.

April, 1905.

Miscellaneous Publication, No. 843.

Farmers' Fowls.

G. BRADSHAW.

CHAPTER I.

Introduction.

WHEN, in 1897, I contributed a series of articles to the *Gazette* on "Profitable Poultry Breeding," and followed with what was intended to be monographs on all the popular breeds of fowls, increased duties at the Export Cold Stores obliged a temporary abandonment of the project, not, however, until the Orpington and Wyandotte breeds were exhaustively dealt with. How the works were received was evidenced by the fact that in a few months the several thousands of reprints were exhausted, and for several years there has been continuous requests for second editions of the respective pamphlets. Anconas have also been dealt with, and the intentions are still good for treating lengthily on all the breeds. However, the rate of progress has been so slow in the past, that to give the histories, adaptabilities, and other claims of the remaining breeds at the same rate of speed as was occupied with those mentioned, long before the last one would be reached other later originations would have appeared, hence no reasonable finality could be expected. To overcome the apparent difficulty of dealing with all, and giving undue preference to none, it has been suggested that as the Orpington and Wyandotte hold premier positions as to popularity, and new varieties of both have originated since the publication of the first articles, that a second series might be undertaken.

This, however, did not get over the difficulty of time, which has been satisfactorily solved by the editor, who, in place of the one-time twenty-five or thirty page contributions, has signified his willingness to accept half that number for each issue, the priority of breeds being as satisfactorily settled by the adoption of the simple title of " Farmers' Fowls " ; and as the agricultural farm is the ideal home for domestic poultry, every known breed and variety can be embraced in the comprehensive title.

In the previous articles, under whatever name, it will be recollected that the chief subjects were the history of the various breeds and their adaptabilities, whether for eggs or meat, or both ; the production of market fowls and eggs, marketing, storing, &c., &c. The present contributions are, however, intended to be more comprehensive, embracing every phase of poultry-keeping, including locality, appliances, runs, coops, breeds, crosses, artificial and natural incubation, feeding, rearing, diseases, fancy and market poultry, &c. ; causes of failure and other disappointments in the industry.

55823   A

The chief departures from the usual writings and teaching will be the absence of any desire to make the subject attractive by unduly magnifying the possibilities of the industry, a lengthened experience testifying to the fact that although some people are making a living from fowls alone, and many largely assisted by the profits resulting from poultry-keeping as a side issue, there are numerous instances, and some late ones too, where goodly sums have been lost in poultry ventures, and to assist in reducing the numbers of these annual failures will be one of the objects of these papers. No promise of a fortune will be held out to those contemplating an investment in the business, for just as with other stock, if correctly incepted in a small way, and intelligently managed, there will be satisfactory progress, and in the course of time a living will be made from it; and if carried on with fruit-growing, or other various issues of the farm, a success can be assured.

Big poultry establishments in this country are very few, and when they do occur they are like the big city businesses—they are not the growth of a day, but the result of small beginnings, a gradual annual growth, interrupted by many unpreventable losses; and when a successful and paying business has been established, the same care and management which contributed to the success has to be as diligently employed, and provision always to be made against possible bad years which every trade, business, profession, and industry experiences. The writer's desire will be to supply the information in as simple a manner as possible, the subject requiring neither scientific phrases nor fine writing to invest it with interest.

Before the close readers will have found some present-day theories and popular prejudices discarded, but whatever the innovations, such have been found by practical experience and recorded in the hope that the novice may derive some assistance in poultry-keeping, and the young fancier help in the way of producing form and feathers. A series of charts, tables, &c., eggs, foods, poultry, prices, &c., new to such works will be given, the whole calculated to show that the breeding of poultry, whether for eggs or meat, or both, is like that of any other business, viz., knowledge, experience, brains, and business methods are essential to, and, in some instances, command success.

CHAPTER II.

Glossary.

In connection with poultry-keeping and breeding, whether for egg-production, carcases for the market, or the fancy, which embraces exhibiting, &c., there are a large number of words and phrases, technical and other terms employed in connection therewith, some of them being quite unintelligible to the novice, and as a number of them may be used during the currency of these articles, the present will be the most suitable place to give a definition of such. The alphabetical order in which they appear making them handy for reference. The chapter on diseases will be reserved till the end of the work.

Addled Eggs.—An egg in which the germ has lived, but from some cause dies and the egg becomes putrid. The word is derived from *adel*, a disease.

Age of Eggs.—This can be fairly accurately told by their density, which decreases as they become older. If 4 oz. of common salt is dissolved in 1¾ pints of water, and a new-laid egg be put into the liquid, it will sink to the bottom. An egg one day old will scarcely touch the bottom, and as the age increases the more will it rise.

Air-cell.—The bubble at the thick end of an egg which supplies the chick with air during incubation. The slightest perforation of the air-cell will prevent hatching.

Allantous.—A bag or sack which develops during hatching, gradually extending until it surrounds the chicken close to the shell, and furnished with a net-work of blood-vessels, and serves as a temporary lung until the chicken is hatched.

Analyses.—The splitting up of foods or other substances into their component parts.

Animal Food.—Is a substitute for the worms and insect life which fowls get in a natural state. Flesh of any animal, cooked or raw, green cut bone, dried blood, or meat meal.

Albumin.—A group of compounds, soluble in water, coagulated by heat. The white of an egg is the chief of this class, and nearly pure.

Artificial Incubation.—The hatching of chickens by any other means than bodily heat. The artificial heat being supplied by oil, gas, or other combustive substances.

Asiatics.—A family of fowls embracing Brahmas, Cochins, Malays, Langshans, and others, all of which lay brown eggs, and are sitters, *i.e.*, become broody.

Aylesbury.—A town in Buckinghamshire, which gives the name to the most popular breed of ducks bred in England.

Barn-door Fowls.—The usual nondescript fowls seen on or about a farm or holding. Sometimes known as dunghills, but more commonly mongrels.

Barred.—Stripes of dark and light across the feathers, as appears in Plymouth Rocks and others.

Balanced ration.—A diet of various foods so arranged or balanced as to get the proper balance of albumenoids and heat-producing foods calculated to give the best results.

Beard.—The muff or bunch of feathers under the throat of some breeds, such as Houdans, Polish, Faverolles, &c.

Breast.—In a dead fowl the under part where the bulk of and best meat is. In a live fowl the breast commences under the throat, and reaches to the keel or breast-bone.

Breast-bone, Crooked.—A bend or twist in the bone, and usually formed when the bird is young, and due to various causes. A serious defect in Game fowls.

Breed.—Any variety of poultry presenting distinct characteristics in type and colour. Type, however, makes the breed, colour the variety, *i.e.*, all Plymouth Rocks should be the same type, whether they are white, buff, or barred.

Brood.—A clutch, flock, or family of chickens under the care of a hen.

Broody.—When a hen shows a desire to incubate she is termed broody or clucking.

Brooder.—An American term for an appliance used in the artificial rearing of chickens ; known in England as a foster-mother.

Breeding in and in.—A system used by the most successful breeder to perpetuate certain traits of type, colour, or other characteristic.

Bone, Green cut.—Fresh bone cut into a fine form, and supplied to the fowls either mixed in the mash or given apart. It is rich in the essentials of egg-production.

Broilers.—Chickens from seven to ten weeks old, and favourite dishes in American hotels. They are split up the back and grilled, the whole bird making a dish for one.

Boned Fowls.—A system of removing every bone from the chicken or fowl, leaving the meat unbroken or torn. An art of the poulterer.

Carriage.—The shape, style, or altitude of a fowl.

Capon.—A male bird deprived of its generative organs, the weight and delicacy being improved thereby.

Caruncles.—Fleshy protuberances to be seen on the head and neck of Muscovy ducks and turkeys.

Chick.—The young of a bird, particularly of a hen, and only applies to a few days old.

Chickens.—In America domestic fowls of any age are termed chickens, *i.e.*, the chicken industry, the chicken market, In England and Australia applied to those weeks or at most a few months old.

Chipped.—An egg is said to be chipped when the young cracks the shell.

Cockerel.—A young cock known as such when under twelve months old.

Comb.—The red protuberance on the top of the male bird's head, taking several forms, such as, single rose, and others.

Condition.—A term usually used in connection with exhibition poultry, referring to beauty of plumage, health.

Crop or Craw.—The bag where the food is stored before it passes to the gizzard to be digested.

Cross-bred.—The progeny from two pure breeds, and when naming the cross that of the male bird is mentioned first. Birds bred from a Brahma cock and Dorking hen are called Brahma-Dorkings, while from a Dorking cock and Brahma hens they are Dorking-Brahmas.

Common Fowls.—The ordinary barn-yard fowls as apart from pure breeds.

Cramming.—A system of artificially feeding young fowls, either by hand or machine, for the purpose of putting on extra flesh, three weeks being the usual course.

Carbo-hydrates.—A food constituent,—starch, sugar, and gum being the leading constituents,—and supplies heat and energy.

Canned Eggs.—A system of egg-preservation in America. The yolks and whites are separated and canned in the same way as meat or vegetables, and in this form are mostly used by cooks and confectioners. The shells are broken up and sold as fertilisers.

Cold Storage.—A system of holding over eggs from a cheap until a dear time, the advantages of the system being the absence of any preservative.

Colour-feeding.—A modern innovation in the poultry world following the boom in buff fowls. The craze for rich colour prompted the administering of cayenne and other pigments during moulting time. A well-known English writer has stated that he knew of many birds too light in colour which were converted into prize winners by colour-feeding.

Crest.—A bunch of feathers on the top of a fowl's head, as the Polish, Houdan, and others.

Cushion.—The rise or mass of feathers on the tail end of a hen's back, largely developed in Brahmas and Cochins.

Chalaze.—Two slightly opaque and twisted cords of dense albumen attached to the yolk and white, and acts as a balancing weight to keep the side of the yolk which carries the germ uppermost.

Deaf-ears.—The skin below the fowl's ears. They are red in the Asiatics, and white in the Mediterranean breeds.

Dubbing.—The removal of the comb, ear-lobes, and wattles, thus leaving the head lean and thin looking.

Dead in Shell.—An egg may be fertile, and the germ develop, but may from lack of strength die at any stage up to the twenty-first day.

Duck-footed.—When a fowl has the hinder toe turned in it is so called, and is a great demerit in Game.

Down.—The finest feathers on ducks, geese, and swans.

Double Mating.—The mating of two different pens of the same breed, the one to produce standard coloured cockerels, and the other to produce pullets of exhibition colour.

Egg-bound.—See Diseases.

Egg-candling.—A system of testing eggs as to quality and soundness, by holding before a strong light in a dark room. Any that show a dark spot, or other trace of unsoundness, being discarded.

Egg-testing.—This is done the fourth or fifth day of incubation, by carefully examining each egg before a strong light. Those that have undergone no change being infertile, and may be removed.

Egg-cooling.—When hatching with incubators cooling for, say, 20 minutes twice daily is essential to a successful hatching, and corresponds with the daily leave the hen takes from her nest.

Egg-turning.—This is usually done twice a day, and in imitation of the hen, who frequently turns her eggs during incubation.

Egg-preserving.—The use or application of vaseline, waterglass, or other substance to prevent decay, and hold the eggs for a rise in market, or other purpose.

Fresh Blood.—The introduction of birds from other yards to increase vigour, or improve the appearance of a flock of fowls, to counteract the effect of inbreeding.

Foreign Eggs.—All the eggs imported into the United Kingdom, from whatever country, are called foreign eggs. These now amount to over six million pounds sterling annually.

Female Influence.—The hen usually influences the internal and vital organs, hence, if a good layer, her progeny may reasonably be expected to be good layers also. There are, however, many exceptions.

Faking.—The trimming, plucking, colouring, or otherwise changing the natural appearance of a fowl, in order to assist it in securing a prize.

Face.—The bare skin round the eyes of a fowl.

Fancy Points.—The breeding for feather, type, and other peculiarities, as apart from economical purposes.

Flights.—The long wing-feathers, called primaries, and used only when flying. They are kept tucked up under the wing.

Feather-eating.—A habit acquired by fowls kept in confinement, idleness usually being the cause.

Fattening.—Feeding fowls liberally to put on an extra quantity of flesh preparatory to killing.

Fluff.—The soft, downy feathers about the thighs of fowls, particularly Brahmas and Cochins.

Furnished.—When a male bird has assumed his full plumage he is known as furnished.

Favus.—A comb disease.

Fatal Defect.—A term applied to features in exhibition poultry which disqualifies them from getting a prize. White feathers in a black fowl, or *vice versâ*. The want of the fifth toe in Dorkings, or the presence of such in a number of other breeds, being an example.

Gapes.—An affection of chickens, due to the presence of worms in the throat. Not prevalent in Australia.

Gout.—Distinct from leg-weakness, by the swelling of joints suddenly.

Germ.—A small disc resting on the top of the yolk, and known as the blastoderm, from which the chicken is developed.

Gargle.—A disease in geese, often fatal among goslings.

Gills.—Sometimes called wattles. The pendulous red skin appendage below the beak, developed largely in the Mediterranean breeds.

Glaire.—The white of an egg—the albumen.

Green Goose.—An English term for a mid-summer gosling which has been principally fed on grass. Another cause for the name is that it is young, *i.e.*, green.

Game Fowls.—The well-known group of fowls, hard in feather, of many colours, and celebrated for fighting propeties.

Grit.—Material such as small stones, broken glass or crockery, shells, and other material eaten by the fowls to grind their food, *i.e.*, fowls' teeth.

Green Food.—Grass, cabbage, lettuce, or other succulent herbage necessary for fowls when penned up, and deprived of their natural food.

Hackle.—The narrow feathers on the neck of a full-plumaged cock. The same sort are found on other parts of the body, and called back and saddle feathers.

Hatching—The incubation of eggs by any means.

Henny Game.—A variety of Game fowls, the male bird being feathered like a hen, having no sickles.

Hen-tailed.—Males with tails like hens, as in the Seabright bantams.

Higgler.—A travelling poultry buyer, who collects for the buyers in Norfolk, Sussex, and Kent, so named by reason of haggling or wrangling over his dealings.

Hen-wife.—A woman who takes care of poultry.

Hovers.—When a hen covers her chickens she is said to hover them.

Hard-feathered.—Game fowls of whatever sort are usually called the hard-feathered varieties.

Hereditary.—Many diseases are hereditary, while defects, such as squirrel tails crooked backs, &c., are frequently inherited.

Hospital.—Quarters on a poultry farm wherein to isolate and doctor sick poultry.

Hydro Incubator.—Incubators whose heat supply is kept up by drawing a quantity of the water away from the tank and supplying its place with some of a higher temperature.

Hybrid.—A cross between domestic fowls and pheasants, &c.

In-breeding.—The breeding together of closely related stock.

Incubation.—The hatching of chickens by whatever means.

Indian Corn.—Maize, a much-used grain for fowls of fattening tendency.

Indian Game. — Well-known fowls, of erratic nomenclature, being of Cornish origin.

Infertile.—A term employed to eggs which have not been fertilised by the male bird.

Influence of Male.—The power of the male bird to implant certain characteristics in his progeny. Propotency.

Insect Food.—The grubs, worms, spiders, and other life which fowls find when on a free range, a substitute having to be supplied when kept in small runs.

Indian Runner.—A rather small duck, of peculiar build ; prolific layers.

Indigestion.—The effect of which is a disinclination for food. See Diseases.

Inflammation.—Frequently brought about by a chill, weakly fowls being most subject to it.

Inheritance.—The likeness inherited from the parents, whether of form, feather, or other feature.

Judging.—The comparison of one bird with another in poultry shows, to determine which have the desired qualities in the greatest proportion.

Joint Disease.—A disease apart from leg-weakness and gout, and incurable.

Jersey Blue.—A fowl of American origin. Its dark legs, however, interfere with its becoming popular. They are of a slaty-blue plumage.

Keel.—Known as the breast-bone, and reaching vertically from the breast in front to the stern. Deep and long is desirable, being then capable of carrying greater quantities of meat.

Killing.—This is done in various ways, breaking the neck and bleeding being the most general.

Knock-kneed.—The hocks being close together instead of well apart.

Legs.—That portion extending from the hock to the foot, and covered with scales. In a table bird the leg is the portion above the hock-joint.

Leg Feathers. — The feathers extending down the outer sides of the legs of the Asiatic breeds.

Leg-weakness.—Young birds of the heavy varieties are frequently troubled with this. If allowed to continue the joint becomes stiff and the toes curled up. There are various causes.

Legs, Scaly.—There are two forms of this disease. The one due to the presence of a small insect, and contagious. The other from a deficiency of oily secretion, causing the scales to dry up and the skin to split.

Lacing.—The edging all round a feather of a different colour, as in the Silver and Gold Wyandotte, Polish, &c.

Laying Breeds. — Leghorns, Andalusians, Minorcas, Spanish, and Anconas are usually known as the laying breeds.

Line Breeding.—A form of scientific inbreeding, the object being to secure and establish certain features in the posterity.

Lobes.—The small skin under the fowl's ear, of round or oval shape. In the Asiatic breeds and their crosses it is red, while in the Mediterranean and Hamburgs it should be pure white.

Liver Disease.—Highly fed poultry in small yards, and denied fresh green food, are subject to hypertrophy or enlargement of the liver. Severe cases are incurable.

Mossy.—Confused, peppery, indistinct appearance in feathers that should be white or other colour.

Marking.—Denoting the colours and peculiarities of fowls according to breed.

Mating.—The selecting of male and female birds to form a breeding-pen.

Manure.—Fowl droppings have a special value in most other countries than Australia as manure. There is a difficulty in its disposal here.

Mallard.—The wild or black duck.

Mash.—The mixture of bran, pollard, and other meals, usually supplied as the morning meal.

Malt Culms.—Sometimes known here as germs, but actually malt sprouts. They have valuable digestive properties, and are rich in fat.

Muff.—A bunch or tuft of feathers under the beak as seen in Houdans, Faverolles, and Polish.

Middlings.—Known also as sharps, pollard, a by-product of wheat and a popular feeding meal.

Moisture.—A term applied in the working of incubators, too little or too much affecting for ill the results.

Musk Ducks. — The well-known Muscovies, a variety not commonly bred outside New South Wales.

Moulting.—The annual dropping of the feathers to make room for the new ones. From the Latin signifying to change.

Mediterranean Breeds. —Anconas, Spanish, Leghorns, Andalusians, Minorcas, all layers of large white eggs, and non-broody.

Modern Game.—The Exhibition or British Game. The chief considerations being colour, hardness of feathers, and length of limb, otherwise known as reach.

Mongrels.—The ordinary fowls seen in city, country, and suburbs, as apart from pure breeds or crosses.

Moubray.—A celebrated writer on poultry about one hundred years ago.

Nest Eggs.—A China or other egg left in the nest as an inducement to the hen to lay.

Neve Crammer.—A modified labour-saving cramming machine on improved principles from the earlier makes of this article.

Negro Fowls.—Another name for the silky fowls, so called from their black skin and bones.

Norfolk Turkeys.—A large black turkey with a few white spots on wing, usually called Norfolk blacks.

Open Judging.—The judging at poultry shows of old was done with closed doors. The demands of exhibitors to get such done in view of the patrons is now largely adopted, and termed open judging.

Ovary.—The cluster of small or rudimentary eggs, resembling a bunch of fruit, found in most adult hens.

Oviduct.—The funnel-shaped egg passage into which the yolk and germ are liberated, and then gets clothed with the white and other element for its development.

Over-crowding.—The keeping of too large a number of fowls in a limited space, thus prompting disease.

Oats.—The best cereal food for laying hens, used more largely in England than this country.

Oatmeal.—In the absence of the prepared foods one of the best foods for young chickens.

Ossifine.—A Sydney preparation of fresh bone in a dry form. A substitute for green cut bone, with better keeping qualities.

Pedigree Breeding.—The breeding of fowls in a scientific manner, as is done with cattle and other stock.

Pullet.—A female fowl in its first year.

Plumage.—The feathering or plume of a fowl.

Poultry-farming.—The keeping of poultry solely as a means of living as distinct from fowls on a farm, where they are but an auxiliary.

Poultry Clubs.—Organisations or societies whose objects are the advancement of poultry-culture.

Portable Houses.—A variety of poultry-houses on wheels, largely used in England, purposed to be removed at intervals on to fresh ground.

Points.—The standards for judging fowls apportions the various sections of the birds into points, the total of which number 100, the purpose being as a guide to judging.

Poulardes.—A pullet deprived of the power of producing eggs, with the object of great size.

Pekins.—The well-known canary-plumaged duck, largely bred on the American mammoth duck farms.

Pin Feathers.—Sometimes known as stubs. The small rudimentary feathers remaining on young fowls or ducklings after plucking, only seen on adult fowls at moulting time.

Prize Poultry.—Pure-bred or fancy poultry whose high breeding entitles them to prizes in a show, but a small proportion of pure-bred poultry are prize poultry.

Partlette.—A name given to a hen with a ruff on neck, a neck apparel worn by ladies in 1600.

Partlette Dame.—A bearded or whiskered hen was so called.

Pea Comb.—The peculiar comb of the Brahma, resembling three combs crushed into one—a triple comb.

Primaries.—The outer flight feathers of the wings, and with which the fowls fly, kept tucked up under the secondaries. The colour of these is important in several breeds when being judged.

Parasites.—The almost numberless sorts of vermin which live on or in fowls. See Diseases.

Pip.—A hard substance on the tongue of a fowl. Not a disease, but the result of disease, analogous to a foul tongue in human beings.

Pulter.—A royal officer who had charge of or to see that the royal household was properly supplied with poultry. The London City Company of Poulterers is still spelt as Pulters Company.

Pheasant Fowls.—Applied in Yorkshire and Lancashire to the coloured Hamburg.

Proteids.—A name given to several constituents of food, as albumen, casein, &c.

Preservation of Eggs.—The application of some coating or element by which eggs can be held over a long time, and retain their food value.

Pencilling.—The crescent-like markings on the feathers of Dark Brahmas, Partridge Cochins, and Partridge Wyandottes. On Hamburgs this peculiarity runs straight across the feathers.

Poult.—A turkey in its first year.

Post-mortem.—The dissecting of fowls, with the object of determining the cause of death.

Prepotency.—The power of the male bird to impress his characteristics on his progeny.

Quantity of Food.—It is impossible to feed fowls by quantity or measure. Breeds differ in their consumption. and the same fowl eats more at one season than at another.

Quality of Eggs.—Northerns, souths. railway, and new laids are Sydney terms for eggs, each having a special quotation, and largely resulting from the age of the eggs, this being assumed from its source of supply.

Rust in Wing.—A reddish-brown colour which frequently appears on the wings of Brown Leghorns—a defect.

Rough in Face.—A term to describe the rough appearance which some Spanish fowls acquire, known as cauliflower face.

Rose-comb.—The well-known broad comb of the Hamburg family, and modified in the Wyandottes.

Repletion.—Some birds over-eat themselves, and move about after in a dejected manner, which is usually the forerunner of disease.

Reversion.—A term used by breeders to describe the occasional throw-back by pure-bred fowls to some original ancestor of, possibly, foreign colour or type.

Roup, Rheumatism, &c.—See Diseases.

Ripening.—The fattening of fowls by artificial feeding. After the usual three weeks' treatment the birds begin to go off, and are then called over-ripe.

Runs.—The usual enclosed spaces in which fowls are kept and bred. When such are of limited extent, and overcrowding takes place, disease assuredly appears. Clean earth, grass, water, and gravel are essentials to health.

Saddle.—That part of the back of a cock reaching to the tail, and covered with sickle-like feathers.

Side-hangers.—The shorter sickle-like feathers on each side of the cock's tail.

Squirrel Tail.—The carrying of the male bird's tail over his back, sometimes touching his head.

Scrabbed Eggs.—A Lenten dish, composed of eggs boiled hard, chopped and mixed with a seasoning of butter, pepper, and salt.

Shaping.—A system in Sussex and other English counties of placing the killed fowls in a V-shaped wooden gutter made of two boards. Weights are placed on the fowls, the object being to give them a more meaty appearance.

Shelter.—In England a shelter-shed is essential to keep off the rain and other inclemencies. In Australia the shed is still more essential, but for an opposite purpose—to keep off the sun.

Secondaries.—The quill feathers of the wings which cover the primaries.

Setting.—The usual dozen or thirteen eggs put under a hen for hatching.

Stag.—A name used for a young cock or turkey.

Spatch Cock.—A term scarcely known in Australia, applied in olden times to a hurried poultry dish. An abbreviation of dispatch.

Sickles.—The long curved feathers of a cock's tail.

Spur.—The offensive sharp weapon on the inside of the legs of adult cocks.

Spangling.—A spot or marking on the end of a feather, best seen in the Silver-spangled Hamburgs.

Slipped Wing.—Frequently the primaries, through some weakness, drop down, appear unsightly, and a great defect in exhibition fowls.

Sex of Eggs.—There are yet some people who profess to be able to tell whether a male or a female will hatch from certain eggs. Every such claim, when put to a practical test, has been disproved. The sex problem, in all life, remains a mystery.

Side Sprigs.—In even the best-bred single-comb birds at times sprigs start from the sides of the combs, which is a serious defect for show purposes.

Spring Chickens.—A term not much employed in Australia. In England the small chicken offered in April and May, and so termed. They realise good prices.

Self Colours.—A one or whole coloured bird.

Shaft.—The thin stem or quill of a feather.

Sharps.—Another name or grade of pollard.

Shank.—That portion of the legs where the scales grow.

Straw Colour.—White fowls, when exposed to the sun, frequently get what is known as a straw colour. This, in most breeds, is a handicap in the show-pen, but in Faverolles is the correct colour for cock's hackle, back, and saddle.

Strain.—A family of fowls possessing some feature in form, feather, or other peculiarity which is transmitted to the progeny.

Standard.—A scale or table of the requirements in shape and colour for the various breeds of poultry.

Symmetry.—Perfection of the various sections of a fowl as distinct from carriage.

Sussex Fowls.—A variety of fowls largely grown in Sussex, of no certain breed, and a variety of colours, but chiefly large-bodied and good market fowls.

Sussex Fowls (dead).—A well-known Leadenhall market term, not necessarily from Sussex, in fact sometimes from Ireland, but specially fattened.

Stolen Nest.—On farms or other places where fowls have a free run they frequently select a secluded laying-place, unknown to the owner, such being called stolen nests.

Soft Eggs.—The shelless egg occasionally found in poultry yards, usually attributed to lack of shell-forming material, but usually the result of the hens being too fat or having ovarian troubles.

Table Poultry.—Fowls bred for, or suitable for breeding market poultry, *i.e.* the table.

Tail Feathers.—The strong stiff feathers of the cock and hen.

Tail Coverts.—The soft curved feathers which cover the tail on both sides. The two top feathers are usually called sickles.

Thighs.—The joint above the knees, and usually covered with feathers.

Tom.—A name given in some English countries to the domestic cock, and in others to the male turkey.

Top Knot.—The bunch of feathers or crest on the top of a fowl's head, as in Polish, Houdans, and others.

Trimming.—The cutting or drawing out of feathers, or other manipulation of a fowl, with the object of improving its chances of winning a prize.

Trussing.—The manner of preparing fowls for roasting, boiling, or grilling, a different system obtaining for each way of cooking.

Tonic.—A medicine whose purport is to give strength and vigour and increase the appetite.

Trio.—A cock or cockerel and two hens or pullets.

Type.—The shape and symmetry of the various breeds of fowls.

Thumb-marked.—A mark or indention frequently appearing on the side of the comb of Leghorn fowls—a defect in the show-pen.

Unfertile, Infertile.—Non-productive. A barren or clear egg.

Undercolour.—The colour of the fowl underneath the surface.

Useful Qualities.—The qualities in fowls which contribute to their profitableness as apart from their appearance.

Variation.—The tendency to reversion or throwing back, well-known to prize poultry breeders.

Variety Classes.—These are for miscellaneous breeds of fowls, which are not provided for in the regular classification.

Vertigo.—An ailment betokening over-feeding to a great extent. A bird so affected will stagger about in a circle.

Vulture Hocks.—Stiff feathers projecting from the hock joints, as seen in Brahma fowls.

Ventilation.—A neglected but most important subject in poultry-house construction. As a rule they are either too draughty or under-ventilated.

Vermin.—Applies to the several sorts of body and feather-eating insects on a fowl, also to those found in nests, houses, &c.

Wasters.—Applied by fanciers to specimens not of sufficient merit to exhibit; culls.

Wattles.—Part of the adornment of the head of the cock, the thin, red, vascular appendage below the beak.

White Comb.—A white, scaly eruption of the comb, named Favus. See Diseases.

Web.—The thin structure of a feather on each side of the stem. Also applied to the skin between the toes of ducks.

Wing Bar.—A band of dark colour across the wing, seen in most parti-coloured fowls.

Wing Butts.—The shoulder of the wing, sometimes called shoulder butts.

Wing Coverts.—The feathers covering the secondaries.

Wings Slipped.—Of frequent occurrence in the Asiatic breeds. The primary feathers appear loose, and hang down below the secondaries; more general in cockerels.

White in Lobe.—In red-lobed birds, such as the Asiatics and Orpingtons and others, white is of occasional occurrence in the red; a defect in the show-pen.

Worms.—Appear in most fowls, and are the cause of a vast number of deaths in chickens.

Wry Tail.—A deformity said to be due to spinal causes; a serious defect.

Whiskers or Beard.—The muff of feathers underneath the beak in Polish, Houdan, Favorelles, and other French breeds.

Waterglass.—Silicate of soda, largely used of late years as an egg preservative.

Wyandotte.—A farmer's fowl, of American origin, whose history will appear in next issue of the *Gazette*, a typical pair of them illustrating this article.

Chapter III.

The Wyandotte.

WHEN, in 1898, I gave the title of "The Wyandotte as a Farmers' Fowl," to a monograph on a breed of fowls then less known than now, its appearance in the *Gazette* prompted complaints as to the nomenclature. The disaffected ones were, however, those who were keeping other breeds of fowls, and rushed to the conclusion that terming Wyandottes a farmers' fowl was tantamount to stating that they were the best fowl for the farmer. It need scarcely be said that no such claim was made for that, or any other breed; my simple contention being that the merits of the Wyandottes, just like the merits of some other breeds, were such as to admirably suit the farmer; that class of the community being in possession of everything considered an essential to the successful and profitable breeding of almost every variety of domestic fowl.

The article throughout never ventured the opinion that Wyandottes were the best fowls for any class of breeders, but was rather a history of the origination of the breed in America, and an exhaustive treatise on its utility merits as testified to by numerous experienced American, English, and Australian breeders, all of whom gave this Yankee origination such a character for the all-round qualities as an egg and meat producer, that to deny the Wyandotte being a farmers' fowl would be to stultify all knowledge of what constitutes the requisites of a breed, suited for those whose object in keeping fowls is profit; and just here it may be said that some people keep fowls whose object is other than that of actual money-making, *i.e.*, as a hobby, just as others do an aviary of canaries, as apart from those who keep one for song; or a cote of fancy pigeons which cost many pounds to purchase and feed, and if exhibited will, perhaps, not win more than a few shillings in the year as prize money. In the same way some people keep Wyandottes; they bring them to the highest standard in shape, colour, and condition, the big outlay frequently yielding no compensation other than the satisfaction of having produced an ideal, which in turn will bring honour and glory to the breeder through the medium of prizes. These people, although frequently losing much on their hobbies, do the industry vast good, and in their absence the Wyandotte and other poultry-yard originations would still be in embryo.

When special articles are written on any breed of fowls, the usual custom is to paint such in colours so attractive that the novice, or would-be beginner, frequently imagines such the best on earth. A lengthened experience, however, with most breeds has shown that, while some have realised far from expectations, others known as the unprofitable sorts have done well, and knowing this I have always been careful, when giving advice, to assert that there is no actual best breed, being well aware of the fact that the profits largely depend on strain, feed, environment, and general management.

Any sort of hens, whether pure, cross-bred, or mongrel, if young and properly cared for, will give a yearly profit of several shillings per head over and above the food bill. At the same time, the best strains of the best laying breeds in existence, through some even minor act of mismanagement, or one of many other causes, may result, and frequently does, in showing a severe loss.

Coming to the Wyandotte. What was written of the breed in the *Gazette* several years ago relative to its merits and prospective popularity, such have, in the extreme sense of the word, been realised. The birds were then in the hands of but a comparatively small number of exhibitors, the patrons being now increased fourfold and more, which can best be confirmed by the numbers then exhibited at the annual shows, and the numbers on exhibition during the past year.

The Royal Agricultural Society's Show will readily be accepted as evidence. In 1899, the show, after my previous monograph appeared, the numbers of this breed exhibited totalled 37, in all the colours. At the same Society's show, in 1904, the numbers had increased to 143, and further than this—excepting a couple of breeders, who have since left the State—the parties who were enamoured with the Wyandottes, and breeders of them in 1898, are patronising them still. This, however, is but a moiety of the evidence in favour of the forward position occupied by Wyandottes amongst the many breeds of fowls. At the above date Wyandottes were only in the hands of a few fanciers, as the following extract from the article referred to will show: "To get an expression of opinion on the merits of the breed from farmers, I determined to communicate with them, but, wonderful to relate, the idea was no sooner entertained than abandoned, from the very fact that to find a farmer in the whole of this colony who bred Wyandottes was an impossibility."

Such a condition of things no longer exists. Any Friday, at the Sydney saleyards of market poultry, Wyandottes, in every stage of growth and adults, are offered in large numbers, these now occupying the place of the mongrels of former times, and although for killing purposes they may not actually fetch much more than mongrels in the same killing condition, still that they are now bred by the poultry and other farmers for market purposes shows that they are more profitable.

Other and important changes have also occurred with the breed in the interim. I refer to the new varieties since originated. Silvers and Golds were the then principal colours, although Whites and a few Buffs had appeared in Sydney at that time.

Mention was made that even Partridge Wyandottes had appeared in England and America, but not in Australia. Blacks, Buff-laced, Violet-laced, Pencilled, and Columbians are all now recognised varieties, the two latter being duplicates in colour of the one-time popular Dark and Light Brahmas. Another and important change has taken place within the past few years in relation to the marking, particularly that of the Silvers. What were considered standard markings or ideals of 1897 would be indifferent specimens now, and fail to catch the judge's eye. The difficulty then was to get clear markings, *i.e.*, the centre of the feather pure white, entirely surrounded with black. There was not much trouble in securing such markings on the breast; but invariably the saddle feathers failed in this respect, the centre being mossy, ticked, or peppery, fluff faulty in colour, and indistinct throughout; while the chief failing in the cocks was sooty hackles and saddle. Another frequent failing in otherwise well-marked birds was a white edge or lacing on the black, known as double-lacing, this putting many a good bird out of competition. Fanciers, however, by careful selection and mating, at last attained the desideratum in many specimens, and no sooner was this done than the fashion was changed, the old-time clear but heavy-laced birds being put aside in favour of still clearer ones, the black lacing being now wanted as narrow as can be, provided no break occur, and this thin line to be as lustrous a black as possible, and known as open-lacing. This change of fashion will be best realised by comparing the pair of Silvers, portrayed in the April *Gazette*, with the ones which illustrated the work in 1898. The Silver hen of that date I described as typical, and connoisseurs in poultry illustrations pronounced it the best that had up to that time appeared. The type and markings were really ahead of anything seen in Wyandotte life; however, as already said, the craze is now for narrow markings, and I think the artist has again produced a pair which, for outline, contour, carriage, shape, or type should be accepted for years to come as an illustrated standard of the Wyandotte breed, while for markings both novice and professional breeder can readily accept them as the recognised fashionable colour-peculiarities of the Silver variety. The above then are the changes which have taken place in appearance since the first publication in the *Gazette* of the Wyandotte as a farmer's fowl.

CHAPTER IV.

Origin.

WITHIN the memory of the majority of those who read these articles, quite a number of new breeds, and very many new varieties of fowls have been created. America has been the seat of origin of most of them, but whether made in America, England, Belgium, or France, and whether of recent or remote ancestry, all except Orpingtons have numerous claimants as the originators; and even this now popular breed, the Wyandotte, although not much more than twenty years since it became recognised as a pure breed, there is much theory as to whom actually laid its foundation, and what breeds formed the first cross; and it must be here recognised that whatever the history of the English Game and Dorkings, the breeds of the present generation are all the result of crosses of the older sort. So far as Wyandottes are concerned, before they received their present name in 1883—which is that of a tribe of American aboriginals—they existed for ten or twelve years previous, but only as a cross; and under a variety of names a great deal of contradictory matter has been published of them.

Perhaps the most authentic history of their origin is that by a well known American poultry journalist, Mr. T. P. M'Grew, and written for the U.S.A. Department of Agriculture, being portion of a report on American breeds of fowls. Mr. McGrew says :—

"The Wyandotte was for years, before it reached its present perfected state, without a name. Its presence was far from attractive, and its average quality was hardly the equal of the common barnyard fowl. So far as the writer remembers, the first Wyandottes were called Sebright Cochins. The result of investigation was convincing that the Sebright Cochin was the product of the union of a Sebright Bantam and a yellow hen, which might have been a Cochin. Such crosses were termed mongrels. While there was little attention paid to them prior to 1870, immediately after that year they began to attract some public notice, and mention was made of them in a few stock papers of New York State. A later investigation has shown that several parties, in the same section of the country, made an effort to produce the Sebright Cochin by crossing the Sebright with the Cochin. This fact is known from letters which passed between those who made the experiment and who interchanged stock, the letters having been presented for publication. The result of the first accidental cross no doubt prompted others to try the experiment. Consequently, the original foundation of what are now called Wyandottes came as an accidental product of an unusual union. The theory of their origin, as accepted by those claiming to be authority, is as follows :—A Mr. John P. Ray, of Hemlock, N.Y., originated a rose-comb fowl by a cross of a Sebright Bantam male and a yellow Chittagong, which he named Sebright Cochins. Others, who became interested (among whom were the Rev. A. S. Baker and Mr. Benson), produced the same kind of fowl. These three gentlemen, became so interested with their newly formed fowls that one of them had them illustrated in the agricultural press during 1872. As a result of the publication of such illustrations these fowls were spread over the country into several States, and were advertised in the columns of poultry journals soon after. Thus, by unguided crosses, was the foundation of this wonderful breed begun. Some carefully planned crosses soon followed, and the able breeder began the labour of moulding them into a set type or form, and of clothing them in a plumage that should be distinctly laced about the edge. To produce the Wyandotte was no inconsiderable task. To bring the solid buff of the original cross into a white centre with black lacing was hardly conceived of in the start, as is proved by the methods employed later. Both Silver-spangled Hamburgs and Dark Brahmas were crossed upon the Sebright Cochin, Silver-spangled Hamburgs and Buff Cochins were bred together, and the best of all these crosses merged into what were called Eurekas; also Excelsiors, Ambrights, American Sebrights, Columbias, &c. While all these many names were applied, as seemed to please the fancy of those working on their advancement, the majority of the fanciers had about concluded to call them American Sebrights, and the managers of the American Poultry Association, at their meeting of 1876, were asked to accept them under this name.

Fortunately for the breed, their admission was refused. This spurred their admirers to more extended efforts, and so, when the time of their recognition came (1883) they were a much improved breed. The name Wyandotte was proposed by Mr. Houdlette, Worcester, Mass., in 1883, and accepted as the future name of the fowl. A general dissatisfaction was shown all over the country at this choice as a name, but those who made the decision should be praised now for their fortunate selection. Time has proved it a most appropriate name, and no one could wish to have it changed.

" The original type of the Wyandotte was the Asiatic, and at the time of origin the standard favoured the Cochin type more than any other. If it were possible to establish these individualities of form, or breed characteristics in the minds of all our poultry people, it would result in our having in our poultry as striking a resemblance of form as we have in our horses, sheep, and cattle. As far as the eye can reach, it is possible to distinguish a flock of merino sheep from any other breed ; the same is true of cattle and horses. The breed characteristics are no stronger in these animals than they should be in fowls. When the Wyandottes were admitted as a breed to a position among our standard-bred fowls, they had reached a form and colour quite distinctive. The males favoured the Dark Brahma in form and colour, the body colour being quite like a well-splashed Dark Brahma. They had smooth legs of a smoky yellow shade, and the rose comb. The females in form favoured our present Silver-spangled Hamburg hens. In colour and markings they were quite crude. Some had greyish white breasts and backs, while others had breasts of white ticked with a darker colour, and backs mossed with the grey of the Dark Brahma. A better description would be that they resembled half-sized Dark Brahmas of very inferior colour and having Hamburg combs and smooth shanks. In many cases the breast feathers of the male were black, with a white stripe through the centre a little larger than the shaft of the feather. The back colour of the male was mixed black and brown, while in the female it was mossed quite like the marking of a very inferior Dark Brahma. Such was the original Wyandotte of this now much valued breed."

CHAPTER V.

Varieties of Wyandottes.

THERE is a peculiarity about the breeds of poultry which does not appertain to other stock of the farm. I refer to the subject of varieties, nearly every breed of fowls being subdivided into varieties, the type in them all being the same, colour and markings forming them into varieties, nor is this subsectioning a later-day development, seeing that the oldest breed we have—Dorkings, for which we are indebted to the Roman conquest,—was moderately early in its history divided into several varieties, the dark or coloured, the silver grey, the red, the cuckoo, and the white, and although now very scarce, some authorities believe the latter to be the original breed.

However, leaving aside the colour of the original Dorking, several of the above varieties have been known for one hundred years and more; and when fanciers at that time, with so few breeds to cross with were able to produce the several varieties or colours, with the distinct Dorking characteristics, it is not surprising that in these later times, and with so many other breeds to work upon, there should now be a multiplicity of Wyandotte varieties. Silver or Black Laced, as the preceding chapter shows, were the first originations, and before it got well established, several breeders were experimenting with the object of a second variety. Gold or Bay Laced was the colour sought; and Mr. Joseph McKeen, of Omro, Wisconsin, is credited with having produced a variety embodying the Wyandotte shape of bay or clay-coloured feathers with a black lacing, and which has evolved into the now well-known markings. It is said that Rose-comb, Brown Leghorns, Partridge Cochins, and Gold Sebrights were the foundation. There were several other claimants as originators, one well accepted statement being to the effect that Indian Game contributed largely to the ground colour of the feathers. Mr. Ira C. Keller, of Prospect, Ohio, one of the foremost fanciers of the breed contributed the following on this colour:—"In 1880 Mr. McKeen crossed the Winnebago fowl with the Silver Wyandotte to produce the Golden. He crossed and recrossed the offspring with the Silver until there was but one-fourth of the Winnebago blood remaining." So the Golden Wyandotte of to-day has but one-fourth to one-eighth of the Winnebago blood left. The Winnebago fowl was a large black-red bird, somewhat the shape of the Wyandotte, with a rose comb, red lobes, and yellow legs. The plumage of the male was much the same as that of the Partridge Cochin, while the hen resembled the Malay Game hen. Other strains were made by raising the larger Wyandotte as the foundation, crossing with Partridge Cochins, Golden Hamburgs, or Brown Leghorns for the desired end.

This variety began to be popular in America in 1886, and as with the Silvers the breed was not long there until the fanciers improved it in both type and markings, and at the end of three years the English people were actually selling these improved Wyandottes back to the Americans for show purposes; the desired rich golden bay in the centre of the feathers supplanting the yellow bay or clay colour, distinctive of the American originations. The only point wherein the Gold Wyandottes differ from the Silvers is in colour, both should be of the same size, type, shape, symmetry, and carriage, the White and Golden Bay constituting the varieties. As with the earlier Silver variety, so with the Gold, the black marking round each feather was much heavier than now, the narrow or open lacing being demanded in both colours, hence the illustration of the Silvers with the above substitution can be adopted as a standard for the Golds as well, and when the Golden Bay is seen in perfection with the lustrous black edging, such constitutes one of the most handsome farmers' fowls extant.

Whites.—It is said the first White Wyandottes originated as sports from the Silvers, and that some of them were known as early as 1882. It is not certain that any Whites were actually made by crossing, as were all the other varieties, or whether all our Whites are bred from the original sports. One thing, however, is certain that the most correctly marked Silvers of the present day occasionally throw white specimens, indeed the Whites come with such persistency that it can safely be attributed to reversion or throwback rather than a sport. It was in 1888 or 1889 that the Whites were admitted to the American standard as a breed, but not till the nineties were they produced in numbers, and now they have become in that country the most numerous of any of the Wyandotte variety, and not only this but have actually eclipsed the Plymouth Rock whether as a breed kept for utility purposes, or by the fancy as an exhibition fowl. The following figures are those recorded at the New York Show in January last, and the three previous years, showing the rise and fall in the estimation of Americans of the breeds which claim American origin :—

	1902.	1903.	1904.	1905.
Barred Plymouth Rocks	228	240	205	265
White Plymouth Rocks	105	78	113	126
Buff Plymouth Rocks	81	79	95	98
Silver Wyandottes	34	28	61	36
Gold Wyandottes	23	29	13	27
Buff Wyandottes	56	48	117	45
Black Wyandottes	11	6	16	19
Partridge Wyandottes	28	41	102	77
Silver-pencilled Wyandottes	...	22	38	42
Columbian Wyandottes	..	16	16	31
White Wyandottes	130	174	232	483

Buffs.—This variety of the Wyandotte family has, like the others, several claimants for the honor of their introduction ; however, "made in America" will suffice, and what helped them along at their initial stage was the fact that buff fowls were then becoming fashionable, and for a number of years they were exhibited in increasing numbers. Many stories were told of this variety as the producers of enormous quantities of eggs, which statements were difficult to understand, seeing that there was a general acknowledgment that the Buff Cochin was largely used in the foundation of the breed. Be that as it may, the last year or two they have not prospered, possibly from the fact that the exhibition colour was so very difficult to breed ; white feathers and black, both disqualifying points, appearing in a large number of the progeny of the best-mated specimens. Several importations of this colour have been brought to Sydney, and although their patrons speak well of the utility qualities, the numbers do not seem to increase.

Partridge Wyandottes.—This variety first appeared in 1889. The colour is a counterpart of the one-time plentiful Cochin of the same name, and are said to have originated from a Gold Wyandotte cock and Partridge Cochin hen. It is within the past three years that this colour has become most plentiful, and during the past show season the highest prices ever given or received for a Wyandotte in any country was

received for a Partridge cockerel of about 8 months old. Quite a
number of high-class specimens have reached Sydney, they appear in
goodly number at the principal shows, and although reputed to be
excellent layers, those tested at the laying competition have comparatively failed as even moderate layers. The breeders of the variety,
however, decline to accept the egg output of the two competing pens
as representative.

In size, type, &c., the Partridge should be like the other
Wyandottes, the colour that of the Partridge Cochin. Although
this variety, like the others, originated in America, there is unquestioned evidence that Mr. Joseph Pettipher, of Banbury, England,
manufactured some of this variety about the same time as was done in
America. Mr. Pettipher made no secrecy of the ingredients, publicly
telling poultry men that he laid the foundation with good Cochin hens.
Since that time the English and American strains have become a good
deal apart in colour, the Americans favouring what is known as a
foxiness in the pencilling of the hens, the male birds being also a
bricky-red colour, as opposed to the rich golden bay of the English
strain.

The first of the American Partridge Wyandottes to reach England
were to the order of Mr. John Wharton, Hawes, Yorkshire, President
of the Wyandotte Club, in 1896, and were first exhibited at Liverpool
Show the same year. The first to reach Sydney were a number brought
out by the late Mr. W. Cook, on his visit here four years ago, the later
arrivals being principally from the yards of Mr. Wharton. Classes are
being now made for this variety in all the Sydney shows, and it is
believed their handsome colour, hardiness, and general good qualities
will contribute to their becoming as plentiful in the immediate future
as any of the other manufactures of the Wyandotte breed. With all
the above colours, the two Sebrights, the Buff, and the Partridge
Cochin engrafted into types of the Wyandotte and called varieties, it
might be supposed that fanciers had a satiety of sorts. Such, however,
was not the case. Wyandottes were booming, and another colour was
again contemplated—that of the one-time plentiful Dark Brahma, or, as
it was once called, Dark-pencilled Brahma. A year or two from the time
of contemplation, this, the fifth variety, was an accomplished fact, and
just as with the other sorts they soon reached England, and, as before,
the breeders there in two or three years had beaten the Americans in
the perfection of pencilling and other markings, and are already sending some of their productions to actually the same State in which they
were originated by Messrs. Ezra Cornell and George Brackenbury.
The latter gentleman was the first to conceive the idea of the Brahma-
coloured Wyandottes, and to produce such mated a dark Brahma hen
with a Partridge Wyandotte cock, and in a number of matings and
other crossings got the desired colour, and this colour being that of
the Dark Brahma in both sexes, the name rightly should have been
Dark-pencilled Wyandottes, rather than this, however, they then
received the name of Silver-pencilled, by which they are still called.
The top-colour of the male is silvery white, hackle and saddle striped
with black, breast and under part glossy black. The hen is a clear

55823   B

grey, pencilled with a darker colour. The type and markings which illustrate this article are considered to be the best that have yet appeared, and will also form black and white illustrations of the Partridge variety previously mentioned. The variety has every promise of becoming plentiful. The first of them to reach Sydney were a number of hens from the yards of Mr. E. G. Wycoff, U.S.A., which took part in the Hawkesbury College third laying competition, and at time of writing, eleven months of this contest has expired, the birds occupying at that time about eightieth place. A few others have arrived to the order of private breeders from England, all of which appear to be of darker markings than the American strains.

Buff-laced.—Another of the sub-varieties, and again made in America, is the Buff-laced, a few of which reached England in 1897. They were, as is always the case, the result of crossing several breeds or varieties with the object of securing a desired colour, or combination of such. The first importations were of very indistinct markings. The English breeder, however, soon perfected the colouring, the Rev. J. Cromblehome being one of the first to secure the desired pattern, viz., a buff feather with a white lacing all around, and in the same manner and uniformity as the black edging on the Silvers. The male birds are marked like the Golds, the breast a rich buff with white edge, the hackle and top feathers striped with white, as the Gold are with black. The colour of the hens is a buff throughout, with a white edging round each feather. Strange to say, this variety has made little headway in the country of its origin, and there will not be surprise if they are allowed to die out there, as the American fanciers appear not to want them, but in England they are taking on well, classes being made for them at some of the big shows. The Countess of Craven is a patron and frequent winner of prizes with this variey, including first at the last Birmingham Show. A few years ago Mr. J. E. Pemell, of Randwick, imported from the originators, Mr. W. C. Keillar, of U.S.A., the only trio of this breed that has yet reached Sydney. The variety being then in its initial stage were as typical as the established sorts, while the colour was indifferent to what was even at that time appearing in America. Mr. Pemell was dissatisfied, and made a present of them to a friend in another State.

Blue or Violet Laced.—These are a counterpart to the above. There are but few of them bred in America, but they are fairly plentiful in England, the only specimens seen in Australia being a few sports thrown by Mr. Pemell's importations. The above two varieties are not likely to be taken up by the fanciers here, while other than delicate markings are essential before they can be recommended as a profitable farmers' fowl.

Blacks.—This colour has been in existence a good many years They have the Wyandotte type in perfection, but so great a difficulty is experienced in breeding the desired yellow legs that the majority of those who took them up have discarded them.

Columbian Wyandottes.—This is the last of the Wyandotte creations, and although most of the others have been a success it is considered that when the Columbian is perfected more varieties will not be wanted. The desired colour in this branch of the family is that of the Light Brahma, and those who recollect the hundreds of pens of the latter breed which appeared in the shows a dozen years ago will at once acknowledge that Wyandottes with the Light Brahma colour cannot be a variety too many. Some forty specimens of them were exhibited at the World's Fair, St. Louis, last year, and it is said they are becoming plentiful throughout many of the American States. In England they have been taken up, and although classes have not yet been made for them, several specimens have secured prizes in the "any other variety" classes. Eighteen months ago a Sydney fancier sent a good sum of money to England for a trio of the breed; the return English mail, however, brought him back his money, with the advice to keep it for a year or two, as the Columbians were not yet good enough to send to Australia. It is said a few of them are at present in Adelaide. There are a few other unrecognised sorts, such as Piles, Cuckoos, &c., but whether as a fanciers' or a farmers' fowl, or both, the established varieties offer selection enough, and the parties who cannot do well by keeping and breeding Wyandottes of any of the above colours need not expect to make a success with any breed of domestic fowls.

Chapter VI.

WYANDOTTES.

Introduction to Australia.

In fancy poultry, of whatever sort or origination, once they become plentiful in England and classes provided for them in the shows, Australian fanciers soon follow suit. In this breed however, they were rather slow, for although specimens reached England between 1868 and 1870, the first of them did not appear here until the tenth annual exhibition of the New South Wales P.P. and D. Society, in 1887. There was but one class made for the breed, and like other fancy poultry then, were shown in pairs. Four entries of them appeared at the show, contributed by Messrs. John Dilling, of Marrickville, C. Irvin, of Burwood, and W. Hope, of Liverpool.

When once introduced, the fanciers took to it as enthusiastically as in other countries. The annual increasing entries at the shows, and the origination of new varieties or colours, warranting additional classification almost every year. The one class with the four entries which appeared the year of their introduction had in a decade increased to ten classes, represented with 150 entries, while the present year's catalogue of the Royal Agricultural Society's Easter Show contains sixteen classes, with an entry of 170.

As already stated, Silver Wyandottes were a good many years established in England before they reached Australia. Not so with the Golds, they having been shown only two years there before they appeared in Sydney, and their introduction here is to be credited to Mr. Hugh Dunlop, of Balmain, who, realising the growing popularity of the Silvers, rightly predicted success for the Golds, and, setting aside the usual methods of importing American breeds from England, went to the fountain head for his stock. The birds were purchased from Messrs. Perrine & Co., large breeders in Alameda, California. The pair cost 30 dollars in America, and nearly as much more in freight and other charges to Sydney, arriving here in August, 1890. The birds were of good type and colour, and equal to any seen for several years afterwards, but a good deal foreign in colour from those which have been arriving from England within the past few years. Shortly after this a pair of hens arrived on the ship "Illawarra," from London. These were purchased by the writer, and afterwards went to the yards of the late T. Hall, of Fairfield. The next importer was Mr. R. Colman, of Bathurst. All the above forming the Gold Wyandotte stock of the State for a number of years.

Since 1898, numerous consignments of this colour have arrived each year from England; a few have also been imported from America, but the latter have been unsatisfactory in both colour and type, and rarely caught the judge's eye when exhibited. In a few years the Golds had overtaken the Silvers in the numbers shown, but of late the latter have again taken the lead. Perhaps the best, if not the most numerous, display witnessed being at the Royal Agricultural Society's 1902 Show, when nine cocks, twelve hens, twenty-one cockerels, and thirty-two pullets were on view. Big displays of the Silvers have also appeared at the fanciers' shows, all tending to the belief that of the laced varieties the Silvers will continue their leading position.

The White variety for a few years did not make much headway, but the last two or three years they appeared to be more plentiful, and exhibited in greater numbers than any of the other colours. There were fifty-two entries of them at the late Royal Show, the next highest being Silvers with forty-three. This is also taking place in America, as is shown earlier in this article, where Whites, at the St. Louis Exhibition, were in numbers several hundreds ahead of any other breed or variety at the exposition. In England also, at the present time, they are more plentifully bred and shown than any of the other varieties.

Buffs were introduced to Sydney eight or nine years ago, and as this colour was then becoming a fashionable one in fowls, it was thought there would be a boom in Buff Wyandottes. Such, however, did not come off. Those wanting a Buff fowl went for that colour in Orpingtons, and up to the present time, although a few good Buff Wyandottes are about, neither the fancier nor the utility breeder appear to want them.

The first Partridge Wyandottes to reach Sydney were brought out by the late Mr. W. Cook, on his visit to Australia about four years ago. For a year or two fanciers were rather shy of the variety; however, the later importations, showing better type and colour, has done much to increase interest in this variety. Perhaps the best to arrive being those which came to the order of Mr. C. H. Bayley during the past year, while consignments for other fanciers by the "Moravian," early in March last, include some of the best that have been bred in England, and should do much to further increase the interest in this variety, which, as the illustration in the *Gazette* shows, is about the handsomest of the whole Wyandotte family.

The Silver-pencilled variety is expected to be exhibited for the first time during the present season. The illustration of the cock of this variety will give an excellent idea of what an exhibition bird should be.

Chapter VII.

Wyandottes for the Table.

Farmers' Fowls being the title to these articles, I now come to the merits of Wyandottes which warrant such an appellation. So far as my own experience goes, I had the honour of placing the first pair of this breed in a Sydney show pen, now nearly eighteen years ago, and have in various ways been connected with, and a patient observer of them until the present day, and, although I do not say that Wyandottes

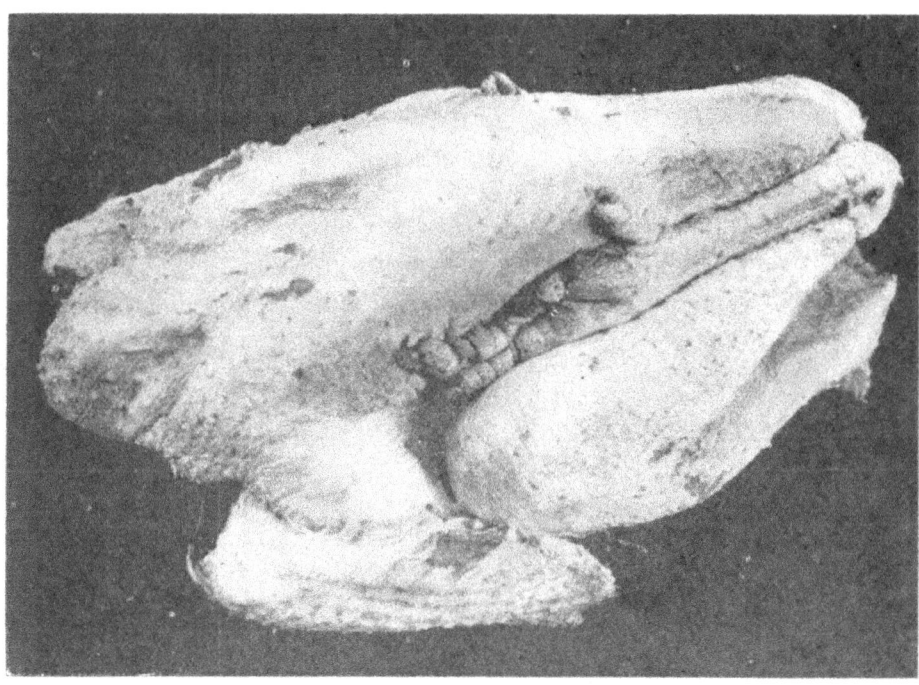

Wyandotte Cockerel.
Hatched, 14th December, 1904 ; killed, 18th April, 1905 ; live weight, 4 lb. 14 oz.

are the best of all, and the only breed, I do maintain they have every essential factor which goes to constitute them a most profitable breed to keep by those for whom these articles are intended, and highly deserving of the highest position amongst the several breeds embodied in the comprehensive nomenclature of farmers' fowls.

One important feature of the breed is the fact that being rather short-legged and cobby in build, with, consequently, finer bones and smaller carcase than the Langshan, Rock, or Orpington, when the latter three breeds, but particularly the first two, are growing a frame-work of bones, on which, later on, to place a big carcase of meat, the Wyandotte is growing flesh and bones at the same time; and if chickens of all the four breeds are in the same flock, it will be found that if caught at any time during the second, third, and fourth month, the Wyandottes will be plumper and fleshier than the other breeds, and always in killing condition within the above period, and, if well fed from hatching to killing time, will require no special fattening for either the local or other markets. After four or five months the other breeds will overtake and pass them in weight; but poultrymen being, like other producers, anxious to realise as quickly as possible,

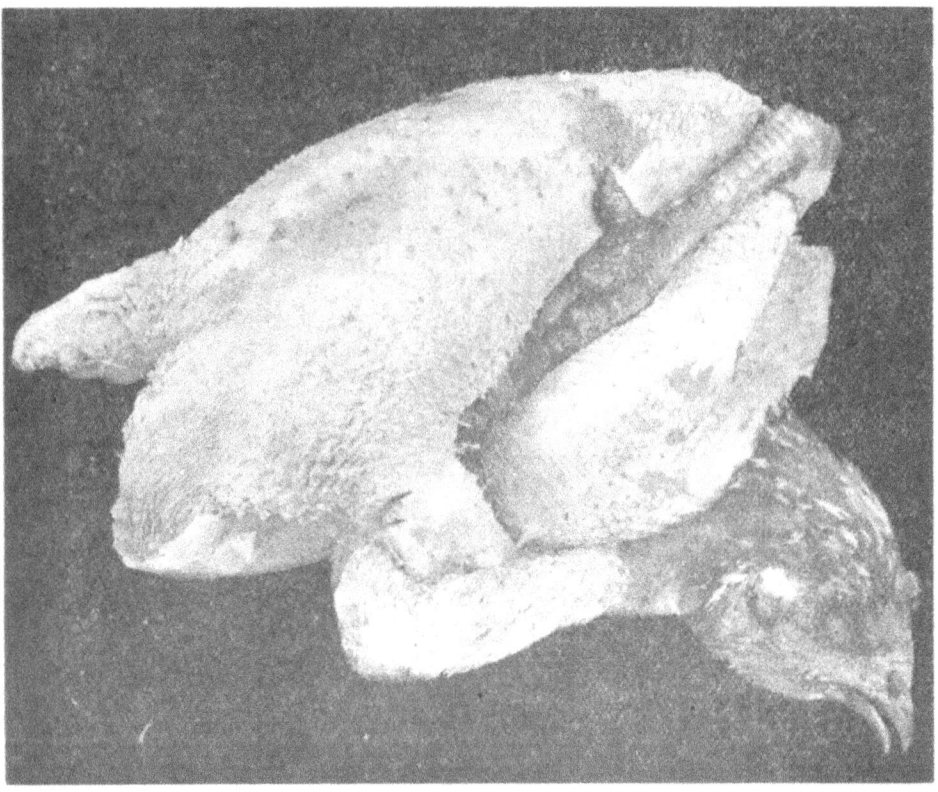

Wyandotte Pullet.
Hatched, 11th January, 1905; killed, 18th April, 1905; live weight, 3 lb. 6 oz.

find it safer to accept 4s. or 5s. a couple for four or five months' old chickens, than to hold over a month or two longer in the mere hope of higher figures.

In my articles on this breed in 1898, I gave comparisons as to the growth made that year with Wyandotte chickens and others of the then favourite Langshan. Since then, I have experimented with several other breeds, and, so far, have had no results better than the Wyandottes, except from a breed with which I am at present experimenting, the results of which will not be available in time for this article.

The following comparison of Wyandotte and Langshans appeared in the *Gazette* in October, 1898. The experiments made with Dorkings and Orpingtons were made two years later, but all were fed and kept approximately alike :—

	Weeks.	Wyandottes. Ounces.	Langshans. Ounces.	Orpingtons. Ounces.	Dorkings. Ounces.
Age	... 4	8	8	8½	7½
,,	... 6	15	15	16	14
,,	... 8	24	22	25	21
,,	.. 10	32	28	29	28
,,	... 12	40	38	38	37
,,	... 14	52	50	48	44
,,	... 16	64	64	65	60
,,	... 18	72	74	75	68

The chickens were carefully weighed each week, but only fortnightly results are given. It will be seen that up to fourteen weeks the Wyandottes made the most weight, the Langshans and Orpingtons then overtaking them, the three breeds scaling 4 lb. for each chicken at sixteen weeks. The Dorkings were a puny lot, and never did well, some of them dying during the course of the experiment, other specimens of the same flock having to be substituted. It should be noted that, although all three breeds give about equal results for the food and attendance, such equality was more apparent than real, for while the sixteen weeks old Wyandottes were plump, fleshy chickens, with good breast, and quite fit to kill, the other breeds being bigger framed and bodied birds were, to this time, but making bone, muscle, and frame to carry the big quantities of meat expected thereon at six or eight months of age.

The accompanying illustration of two chickens of this breed (cockerel and pullet) further emphasises the contentions that whatever the merits of other breeds or varieties of fowls may be, the Wyandotte is one that cannot be overlooked where fast-growing plump carcases are wanted, this feature alone constituting them a farmer's fowl.

The cockerel was hatched on the 14th December and the pullet on 11th January, and both killed on 18th April, and photographed by Mr. Grosse, artist to the Department of Agriculture, on the same day. The weights are better than those tabulated above, the cockerel weighing 4 lb. 14 oz. at eighteen weeks, and the pullet 3 lb. 6 oz. at fourteen weeks, live weight.

With this and the previous records as to meat capabilities of fowls, those desiring such essentials need have no hesitancy in adopting any variety of the now favourite and plentiful Wyandotte.

CHAPTER VIII.

Wyandottes for Eggs.

IT has been already shown that owing to the comparative scarcity of this breed of fowls in New South Wales eight or more years ago, there was much difficulty in securing reliable testimony as to the merits of the breed, for although my experience with them was most satisfactory, I was too well acquainted with the erratic nature of the domestic hen to accept my own experiences as conclusive evidence of the characteristics of the then Wyandotte; and I may be here permitted to remark that the man or woman who finds the breed of fowls which he or she keeps satisfactory, there should be much caution before changing for an expected better one; and if a patron of some breed or variety finds such unprofitable, and a determination made to try another, then when electing this new one, a good character of it should be obtained from more than one source, seeing that in egg-production, strains or families play an important part; some Wyandottes being just as bad performers as are to be found in a flock of common fowls, and on this subject of strain the English Utility Poultry Club have at each of eight laying competitions attached the following note to its reports :—" The competitions are not given to determine which is the best breed, the club recognising that good laying is a question of strain and not of breed."

Realising all this, and in order to secure an opinion from the patrons of Wyandottes as to the merits of this breed as then known, I communicated with the most prominent of them in this State in the following terms :—

(1.) What length of time have you kept this breed?
(2.) Are the chickens easier or more difficult to rear than other breeds?
(3.) At what age do the pullets begin to lay?
(4.) Do they grow as fast and develop as quickly as other breeds?
(5.) What is your opinion of them as farmers' fowls?
(6.) What is your experience of them as layers?

One breeder wrote :—" Having a very limited space in which to keep fowls, I tried nearly all the popular breeds, and can safely say I was never so well suited as I am by Gold Wyandottes for good utility fowls; having such remarkably small wings they do not attempt to fly, are quiet and contented in a small space, and are not large eaters. They lay quite as well as any of the best-laying breeds, while as table fowls they are second to none, a combination hard to produce, and as sitters and mothers are excellent."

The proprietor of a long-established poultry farm wrote as follows :—
" I have kept Silvers and Golds about eight years, and Whites two years. I find they grow and develop quickly. I have had cockerels weighing 5½ lb. at five months old. My winning Golden cockerel at the last show was eight months old and weighed 8 lb. The chickens are easy to rear. I have more successful hatchings of Rocks and Wyandottes than any other variety. In mixed sittings of Rock and Wyandottes I see no appreciable difference in growth and development of chickens; the pullets occasionally lay at five and six months old. I am not in favour of forcing pullets to lay too soon, and try to keep them back. I believe that laying before seven months affects or stops the growth of the hen. My experience is that early laying pullets make the poorest layers as hens; all the Wyandottes are good winter layers; you can rely on them for eggs when prices are high. They

make a good cross, but no cross that I ever kept can equal the pure Wyandotte, either for weight or quality of flesh. They are good sitters and mothers, and are easily broken off being broody. The demand for stock birds and settings of eggs is steadily increasing in all varieties—all are so good, I cannot say which I prefer."

Another firm of breeders and exhibitors contributed the following:—"We took up Wyandottes—Silver-laced—about seven or eight years ago, and soon after the Gold variety. Since then we have added the Whites to our stock. We find the chickens very hardy, and more easily reared than other breeds; they grow fast and develop very quickly into fine blocky birds. We have had them lay between four and five months old, and as layers they compare favourably with either Leghorns or Minorcas, and as winter layers they beat either of those two varieties; while as farmers' fowls we find them as near perfection as it is possible to get any breed, for the reason that they are good layers, good birds for the table, good sitters and brooders, very quiet and docile, while a 4-foot fence will confine them."

Another breeder's contribution was:—"I have kept the Wyandottes since shortly after their arrival in the colony. I have found the chickens very hardy and easy to rear, and grow quicker than any other variety. The pullets commence to lay at about five months. I find them first-class layers, and as farmers' and general purpose fowls, none can surpass them."

A further contribution reads:—"I have kept the Wyandottes for about five years, and have found the chickens quite as easy to rear as any of the other breeds. Some begin to lay at five months, others not until seven or eight months. The chickens grow quickly, but feather a bit slowly in cold weather. As farmers' fowls, I consider them next best to the Orpington; the latter I consider the king of utility fowls, especially the Buffs. The Wyandottes are first rate layers all the year round; I could not wish for better. The sale of stock-birds and eggs is not so good as that of other breeds I keep, but the public are beginning to find out what a grand fowl it is, and I predict a great market for it at an early date."

A large poultry breeder wrote:—"I find the chickens the easiest to rear of any breed I have handled up to the age of twelve weeks. They are the fastest growers of any breed I rear, and are exceedingly plump and well developed. I have had pullets start to lay as early as four and a half months; but they usually commence at five and a half or six months. As layers, they are the best of the sitting breeds, especially as winter layers. I consider them the ideal fowl for farmers, and always recommend them as such, for they lay well when eggs are dear, which also means early sitters, thus enabling the farmers to get early crops of chickens, which can be marketed at a price more than double that of later hatches. They are easily confined, and always ready for market, whether at three months or three years."

The last contribution was from an experienced breeder, who wrote well of the breed as quick growers, &c., but found their performances as layers far from good, the annual average being from 110 to 130 eggs per hen per year.

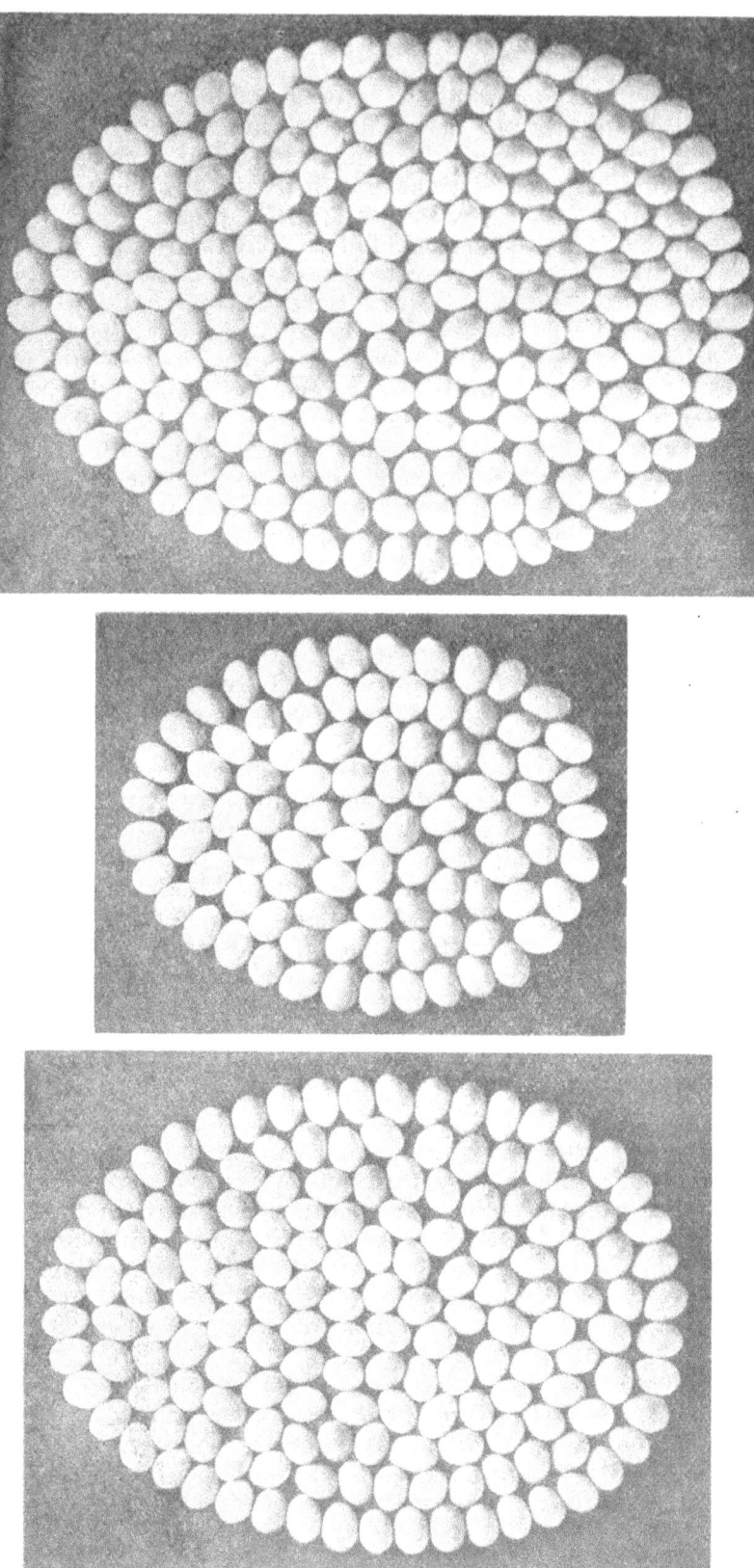

145 Eggs. Average laying of 186 Wyandottes for twelve   89 Eggs. The lowest record made in twelve   204 Eggs. The highest record made in twelve months.
months.   months.

From records of the Third Egg-laying Competition at the Hawkesbury Agricultural College, 1904-5.

The above opinions were given seven years ago, and despite all the newer breeds and varieties which have since originated, the majority of the contributors continue breeding the Wyandottes, which is con-clusive that their lengthened experience has confirmed their earlier convictions that the breed for all practical purposes, as the Americans say, fills the bill. It will be noted that one of the contributors rather discounted the egg-producing merits of the breed, from his experience of them as poor performers at that time, there are a proportionate number of people with the same belief at the present day, and all confirmatory of my frequent contention that breed in fowls is but one thing; strain playing a larger part in the profitableness of poultry keeping when eggs are the principal object.

It may be here thought, that in order to secure more up-to-date opinions about egg-production than that supplied some years ago, the present patrons of the breed might have been consulted. What, however, was practicable at that time is not so now, owing to the extraordinary increase in the number of breeders, for in every town, suburb, and settled district in the country Wyandottes abound. Still the records have been obtained, and in a more unquestioned manner than any other available way. I refer to the English laying compe-tition, and that conducted at the Hawkesbury Agricultural College.

The English laying tests were incepted by the Utility Poultry Club of that country, and are held only throughout the severe winter months of each year, with the object of encouraging laying when eggs are at their dearest. The first competition was held in 1896, and repeated each winter, and of the eight now held, four of them have been won by Wyandottes; that which concluded on the 22nd of January last was won by Leghorns, a breed not reputed as great winter performers, this fact again bringing into prominence the uncer-tainty of hens no matter what breed.

In connection with the last English competition, each hen's laying is recorded by the now much adopted trap nest, its advantages being, that rather than the produce of the four birds being averaged at the close of the test, and the bad layers in the pen getting credit for as many as the good layer, the recording nest system admits of no aver-aging, each hen getting credit for her actual performance.

The manager of the English competition points out the importance of this feature, by showing that while the winning pen averaged 61 eggs each hen in the sixteen weeks, the actual laying, individually, was as high as 69, and as low as 49. Again, while the third pen of White Wyandottes had an average of 57 each in the sixteen weeks, the trap nest actually showed that one of the hens in this pen made the record for the competition of 73 eggs, while another one in the same pen only produced 37. Consequently, had the trap nest not been in use, the latter drone would have got credit for 57, viz., 20 eggs more than she actually laid, while the hen that put up the record of 73, by the average system would have been credited with but 57, an actual loss of 16 eggs. However, taking the last competition all in all, of the 36 pens, Wyandottes at the close occupied third, fourth, sixth, eighth, ninth, tenth, and some lower places, and, just like more than one of the tests here, a pen of the same breed came in last : and again—thanks to the recording nest—two hens (both White Wyandottes) did not lay an egg during the course of the competition. In the absence of this valuable invention the two specimens, under the average plan, would have got credit for 34 and 22 eggs respectively.

Coming to the Hawkesbury competition for evidence in favour of the laying properties of Wyandottes, the testimony will be genuinely accepted here. The birds were drawn from all parts of this State, some from Victoria and America, owned by fanciers, poultry-farmers, and other breeders. They were all young birds, under the same control, fed, housed, and otherwise managed alike, and faithful records taken, hence the figures for the twelve months' test will stand as testimony in favour, or otherwise, of the many breeds which took part therein. The final result was again overwhelming evidence that there are good and bad layers in all breeds, and particularly so in Wyandottes. The winning-pen of six birds of the silver variety finished with 1,224 eggs to their credit, being an average of 204 for each hen, while, to complete the disparity, another pen of the same breed laid but 533, or about 89 for each hen, considerably less than half the number laid by the winners. It may be retorted by some, that although the highest record was made by the silver-laced variety, the pen that did so badly, although Wyandottes, were of a different variety. This contention is quite true. However, there is not much in the admission, seeing that the Silvers were lowest, but one, i.e., occupying 99th place at the final. The illustrations accompanying this article will best show the difference between the good and bad layers. The larger photograph shows the actual number of eggs, 204, laid by each of the hens in the winning-pen, the smallest picture showing 89 eggs, the output of the lowest. It is, however, with the centre figure that poultry-keepers will be most concerned, as it affords the safest basis on which to approximate the egg expectation from an average flock of young Wyandottes during their first year's laying, and should this flock be kept two, three, or four years more, the yield gradually diminishes, few hens during their third year producing more than pays their feed-bill.

Of the one hundred pens competing at the college, thirty-one lots were Wyandottes, representing the different varieties, as follows :— 17 Silvers, 6 Whites, 3 Golds, 2 Buffs, 2 Partridges, and 1 Silver-pencilled, totalling 186 birds. The total egg-production for the entire lot for the twelve months averaged 145 eggs for each hen, this being the number represented in the centre illustration. In other words, a flock of young healthy Wyandottes of the several varieties, if well fed and otherwise consistently managed, may be expected to lay twelve dozen eggs each, during the first twelve months of their performance.

The reliability of the Hawkesbury test, for the purpose of this article, has been above briefly referred to. The other varied lessons to be learned from it will, I feel sure, be dealt with by Mr. Thompson, the Poultry Expert, Hawkesbury Agricultura' College.

CHAPTER IX.

Judging Wyandottes.

In judging poultry at the various shows in this country, what is known as the English Poultry Club's standard is adopted, as opposed to the American standard, which in every breed is in many essentials much different.

The following extract, which forms a portion of the introduction to the club's standard, will assist in the proper application of the compilation :—"When three or four good judges of any particular variety of fowls assemble before a few good specimens of it, it will generally be found that they agree in their conclusions as to which is first, second, and third. There are, of course, exceptions ; there may be some strong personal interest, and some people have a specially favourite "point," and there are occasional cases of real doubt, when it is very difficult, balancing one point against another, to decide which really is the best in one point, and some other in another, none being alike good in all. As there are many points to be taken account of in every fowl, such a simple fact as this shows that there is some proportion generally accepted, however tacitly and roughly, between the judging value of those points, or of defects in them. For the general opinion, united in as above, does not depend upon the best bird being the best in any one cardinal point; otherwise a class of twenty could be judged in ten minutes. However insensibly and informally it may be done, the aggregate of points or defects have to be weighed, and it is acknowledged by all that excellence or defect in some points is not of so much importance as in others. The standard thus prepared and presented aims first at correctly describing the varieties treated of in language as simple and comprehensive as possible. In the second place it desires to lay down the fair proportionate value which general opinion considers should be given to any defects in the various points. These proportionate values thus arrived at, it is hoped that qualified judges will recognise and respect, not violently upsetting them by notions of their own; though it is not likely, nor perhaps desirable, that birds should be systematically "scored" by them and prizes awarded accordingly. This has been done in America for years; but it is becoming more doubtful if the system will continue, the larger shows being now judged otherwise. The proper use of a standard is not to give birds a score, but to place them in correct order of merit. It must never be forgotten that small deductions or cuts for conspicuous defects cannot do this. The figures in the "points" following are meant to express what ought to be deducted from the standard 100 points, for as much fault in the points named as can exist, and still leave the bird in competition. Not as much as possible, by any means: for instance, if the point be comb, and ten are allowed, a comb bad beyond a certain degree would throw a bird entirely out, and not be deducted at all. It is meant that if the comb is really about as bad as still leaves the bird any chance at all, the ten should be deducted ; and less for slighter defects, perhaps even only one or half a point for very slight defects. But for serious and evident faults *serious cuts must be made* if the standard is to perform its function. Though not employed in systematic judging, it is suggested that in cases of doubt—which sometimes arise, and are then too often decided haphazard, or by some preference—the standard might be applied.

Wyandotte Standard.

General characteristics of cock :—

Head and neck.—Head : Short and broad.

Comb.—Rose, firm and even on head ; full of fine work ; low and square at front, tapering towards the spike, which should follow the curve of the neck.

Face.—Smooth and fine in texture.

Ear-lobes and wattles.—Medium length, fine in texture.

Neck.—Medium length, well arched, with full hackle.

Body.—Breast : Full and round, keel bone straight.

Back—Broad and short.

Saddle.—Full and broad, rising with concave sweep to tail.

Wings.—Medium size, nicely folded to the side.

Tail.—Well developed, spread at base ; the true tail feathers carried rather upright ; sickles of medium length.

Legs and feet.—Thighs : Of medium length, well covered with soft and webless feathers.

Fluff. — Full and abundant.

Shanks.—Medium length, strong, but fine in bone.

Toes.—Four in number, straight, and well spread.

General shape and carriage.—Graceful and well balanced, resembling a Brahma.

Size and weight.—Rather large. Matured cockerels, about 6½ lb. for Buff-laced, 7 lb. other colours ; adult cocks, about 7½ lb. in Buff-laced, 8½ lb. in other colours.

General characteristics of hen :—

Head and neck.—Head, comb, face, ear-lobes, and wattles, as in the cock, but the appendages smaller.

Neck.—Medium length, with short, full hackle.

Body.—To correspond with that of the cock.

Back.—Short and wide at shoulder.

Tail.—Well spread at base.

Legs and feet.—As in the cock.

General shape and carriage.—To correspond with the cock.

Size and weight.—Pullets, 5½ lb. in Buff-laced, 6 lb. in other colours ; adult hens, 6½ lb. in Buff-laced, 7 lb. in other colours.

Colour of Silver Wyandottes :—

In both sexes.—Beak : Horn colour, shading into or tipped with yellow.

Eye.—Bright bay.

Comb, face, ear-lobes, and wattles.—Bright red.

Shanks and feet.—Bright yellow.

In the cock.—Head : Silvery white.

Neck.—Silvery white, with clear black stripe through centre of each feather ; free from ticks.

Saddle.—Hackles to match the neck.

Back.—Silvery white, free from yellow or straw colour.

Shoulder tip.—White, laced with black.

Wing-bow.—Silvery white.

Wing coverts.—Evenly laced, forming (at least) two well-defined bars.

Secondaries.—Black on inner and wide white stripe on outer web, the edge laced with black.

Primaries or flights.—Black on inner web, and broadly laced white on outer edge.

Breast and under-parts.—The web white, with well-defined jet black lacing, free from double or white outer lacing, the lacing regular from throat to back of thighs, showing green lustre.

Under colour.—Dark slate.

Tail.—True tail feathers, sickles, and coverts, black, showing green lustre.

Thighs and fluff.—Black or dark slate, powdered with dark grey, with clear lacing round hocks and outer side of thighs.

In the hen.—Head : Silvery whie.

Neck.—Silvery white, with clea black stripe through centre of each feather, free from ticks.

Breast and back.—Under-colour, dark slate ; web, white, with regular, well-defined jet black lacing, free from double or outer lacing, and showing green lustre.

Wings.—Same as back, on the broad portion.

Secondaries and primiries.—As in the cock.

Tail.—Black, showing green lustre ; the coverts black, with a white centre to each feather.

Thighs and fluff.—Black or dark slate, powdered with dark grey.

[N.B.—Regularity of lacing and quality of colour in all cases to count above any particular breadth of lacing.]

Colour of Golden Wyandottes :—

In both sexes.—Beak : Horn colour, shading into or tipped with yellow.

Eye.—Bright bay.

Comb, face, ear-lobes, and wattles.—Bright red.

Shanks and feet.—Bright yellow.

In the cock.—Head : Rich golden bay.

Neck hackle.—Rich golden bay, with distinct black stripe down the centre of each feather, free from ticks, black outer edging on black tips.

Saddle hackles.—To match the neck.

Back.—Rich bay, free from black or from deep maroon.

Breast and wings.—Same as Silvers, substituting rich golden bay for white in ground colour.

Thighs and fluff.—Black or dark slate slightly powered with gold, with clear lacing round hocks and outer side of thighs.

[N.B.—Brightness and uniformity of colour to be considered of more value than any particular shade.]

In the hen.—Head : Rich golden bay, with distinct black stripe down the centre of each feather, free from ticks, black outer edging, or black tips.

Breast, back, wings, and tail.—As in Silvers, substituting rich golden bay for white as the ground colour.

Thighs and fluff.—Black or dark slate, slightly powdered with gold.

[N.B.—Brightness and equality of ground colour, and regularity of lacing throughout, to be of first importance.]

Colour of White Wyandottes :—

In both sexes.—Beak : Bright yellow.

Eye.—Bright bay.

Comb, face, ear-lobes, and wattles.—Bright red.

Plumage.—Pure white throughout ; straw-colour to be avoided.

Legs and feet.—Bright yellow.

Colour of Buff Wyandottes :—

In both sexes.—Beak : Rich yellow.

Eye.-- Bright bay.

Comb, face, ear lobes, and wattles.—Bright red.

Plumage.—Any shade of buff from lemon buff to rich buff, on the one side avoiding washiness, and one the other side a reddish tinge. The colour uniform throughout, allowing for greater lustre on the hackle, saddle, and wing-bow, in the case of the cock only.

Legs and feet.—Bright yellow ; toe-nails white.

Colour of Partridge Wyandottes :—

In both sexes.—Beak : Horn colour, shading into or tipped with yellow.

Eye.—Bright bay.

Comb, face, ear-lobes, and wattles.—Bright red.

Legs and feet.—Bright yellow ; toe-nails horn colour.

In the cock.—Head : Rich orange.

Neck.—Orange or golden red, with paler shade at back, each feather having a glossy black stripe down centre.

Back.—Rich dark red, free from maroon or purple shade.

Saddle.—As in the neck hackle.

Wings.—Rich red, as in back.

Wing-bar.—Solid black.

Secondaries.—Rich bay on outer web, and black on inner web and end of feather, the rich bay alone showing when the wing is closed.

Breast.—Black, free from ticks.

Fluff.—Solid black.

Tail (including sickles and tail coverts).—Glossy metallic black.

In the hen.—*Head* : Rich orange.

Neck.—Golden yellow, striped with black.

Breast, back, and wings.—A light brown ground-colour, free from red or yellow tinge, every feather distinctly and plentifully pencilled with a darker shade. Pencilling uniform throughout, to follow the form of the feather. A brick or yellow ground colour objectionable.

Fluff.—Brown (free from yellow or red), slightly pencilled (the more pencilled the better.

Tail.—True tail feathers black, shading to brown at top, which should be well pencilled.

Colour of Silver-pencilled Wyandottes :—

In the cock.—*Head* : Silvery white. *Neck hackle* : Silvery white, sharply striped with black in the centre of each feather. *Breast* : Glossy black.

Underpart of body, thighs, and fluff.—Black.

Back and shoulder coverts.—Silvery white, except between the shoulders, where the feathers should be black laced with white.

Saddle.—Silvery white, striped sharply with glossy black.

Wing-bows.—Silvery white.

Tail.—Black.

Legs and feet.—Bright yellow.

In the hen.—*Head* : Silvery white.

Neck hackle.—Silvery white, striped with black.

Tail.—Black, or black edged with grey, or with pencilling.

Remainder of plumage.—Ground colour any shade of grey, but not too brown, with a pencilling of black or a darker shade of grey than the body colour, very clearly defined, following the outline of each feather, as uniform in character as possible ; the pencilling or bands on each feather to be numerous.

Legs and feet —As in cock.

Value of Points in Wyandottes.

Gold or Silver.—Cock or hen.

Defects.	Deduct up to
Defects in comb	8
,, head	5
,, ear-lobes and wattles ...	6
,, neck	8
,, breast	14
,, back	14
,, tail	7
,, wings...	12
,, fluff	6
,, legs	6
Want of size and condition ...	14
A perfect bird to count ...	100

Buffs.—Cock or hen.

Defects.	Deduct up to
Defects in comb	8
,, head	5
,, ear-lobes and wattles ...	8
,, neck	4
,, breast	5
,, back	6
,, wings...	5
,, tail	5
,, fluff	4
,, colour	30
,, legs	6
Want of size and condition	14
A perfect bird to count ...	100

Whites.—Cock or hen.

Defects.	Deduct up to
Defects in comb	8
,, head	6
,, ear-lobes and wattles ...	8
,, neck	10
,, back	10
,, body	12
,, wings...	10
,, tail	8
,, legs	8
Want of size and condition ...	20
A perfect bird to count ...	100

Partridge.—The cock.

Defects.	Deduct up to
Defects in comb	8
,, head	5
,, ear-lobes and wattles ...	6
,, neck	12
,, breast	10
,, back	12
,, wings	10
,, tail	7
,, fluff	8
,, legs	8
Want of size and condition	14
A perfect bird to count ...	100

Partridge.—The hen.

Defects.				Deduct up to
Defects in comb ...	...	...	...	8
,, head ...	...	...	...	5
,, ear-lobes and wattles			...	6
,, neck ...	...	...	...	12
,, breast	...	...	...	10
,, back ...	...	...	...	12

Defects in wings...	...	...	...	10
,, tail ...	...	...	...	7
,, fluff ...	..	...	...	8
,, legs ...	...	...	...	8
Want of size and condition	...		...	14
A perfect bird to count			...	100

Serious defects, for which birds should be passed.—Any feathers on shanks or toes; permanent white or yellow in ear-lobes, covering more than one-third of their surface; combs other than rose, or falling over on one side, or so large as to obstruct the sight; wry tails; deformed beaks; crooked backs; shanks other than yellow in colour (except adult cocks and hens, which may shade to light straw colour); feathers other than white in Whites; white in tail, or any conspicuous spotting or peppering on ground of the feathers in Silvers or Golds.

CHAPTER X.
Breeding Wyandottes.

WHEN introducing the present series of articles, it was mentioned that during their course there would be some departures from the beaten track of poultry literature, and already this has been done, the last chapter which gave the standard for judging being one of these departures.

Hitherto poultry-books dealing with the subject of judging mysteriously appended the standard at the end of all the descriptive matter about the breed. An instance may be given of the breed now under consideration—Wyandottes. The works in question correctly tell us how to secure the yellow legs, the rose comb, the red ear, cobby build and colour, and after all this conclude their work by giving the standard which says the legs must be yellow, the lobes red, the build cobby, &c. It will be apparent to the merest tyro that the description of a breed, *i.e.*, the standard which tells all about what the shape, colour, and peculiarities of a breed are, should rightly appear early in the work, the writer following with information as to how these features can be secured.

In now devolves how best to produce the shape, various colours, and other characteristics demanded in the standard for the many varieties of Wyandottes. Of course, it may, or can be said, with reason, that in farmers' fowls, that is fowls best suited for the farmer, it is of little account that the shape or colour conforms to some, perhaps arbitrary, provision stated in some book or demanded by poultry judges; that eggs and meat are not dependant on the colour of a fowl's beak, or whether its legs be bright yellow or a pale primrose colour. This is largely true, but only so in a certain sense, for as already shown many farmers have discarded the too common hen in favour of pure breeds, and should any of these breeds be Wyandottes, it is not sufficient to secure a stock of such, and allow them to breed at their own sweet will, as was allowed with the common sort, for, if so, the progeny in a very short time would, so far as appearance go, be of a varied nature indeed.

55823   C

Wyandottes, as has been shown, are a comparatively new breed, in fact they are described by Harrison Weir and some of the old writers as crossbreds. They certainly have been produced from other pure breeds, and, by scientific mating and selection, will produce their kind, but only in a kind of way as will be shown.

A farmer purchasing, say, the best first prize cock at the Royal Show illustrated in this number, and the first prize hen of the same breed, would naturally expect to breed from them a large number equal to their parents in appearance, and go on breeding as was done with the mongrel sort, and continue increasing his Wyandotte flock, and retaining the colour and type of the variety. Such is not so. From this pair of birds as above, if unrelated, it is just possible that out of the first fifty of the progeny there might not be a single specimen

Mr. W. H. Lathlean's imported Silver Wyandotte Cock.
Winner First Prize at Royal Agricultural Show, 1905.

equal to either parents in lacing or other colour characteristics. But even if there were a few as good in appearance as those they were bred from, if allowed to run and breed together without selection by the owner, the colour and markings of the entire flock, in a very few years would possess little of its former beauty, and in appearance little better than that of ordinary fowls. In process of time a single comb would appear on some specimen, the result of a throwback to an early ancestor—this feature would be reproduced in increasing numbers; feathered legs from the same cause would be in evidence, and would become still more pronounced as time went on; the markings each year would be more diffused, and before many years the entire Wyandotte stock would be that in name only. During the time that this process of deterioration was going on the same cause would be contributory to a diminishing egg-yield and all showing that the simple fact of purchasing Wyandotte stock of the standard quality is insufficient for the owner whose object is meat and eggs, for it must be remembered that the farmer or other breeder, who keeps and correctly mates and breeds pure poultry of any breed, will always be able to dispose of a quantity of eggs for hatching purposes, at considerably more than the ordinary market value, hence a knowledge of the peculiarities of the strains of his stock and experience in mating will prove principal factors in keeping a stock of fowls not only pure-bred, but pure-looking, and considerably more profitable withal.

The silver variety of Wyandottes, being the first originated, it might readily be accepted that twenty or more years should have been sufficient to have established the colour or markings in the variety to an extent embodying the old principle of like producing like. This certainly has been accomplished to an extent, seeing that most of our present-day Wyandottes will produce stock equal, or better, than those existing twenty years ago. This, however, does not satisfy fanciers, who are always ambitious of producing higher ideals than those existent, and the extent to which this has been accomplished will be best realised by the fact that the birds which won first prizes in the Sydney shows a few years ago would be well rewarded with but a commended card if shown at the present day. Smoky hackle feathers, and rusty feathers on back, too white or too black on breast, was then of common occurence in the male birds, while the hens were badly laced, and much trouble in getting specimens free from the objectionable mossy or peppery saddle feathers. Of recent years these faults are scarcely in evidence, many of the birds now exhibited being close up to the standard requirements. The standard tells us that the cock's beak should be horn colour, shading into or tipped with yellow, the eye bright bay, and the comb, face, ear-lobe, and wattles bright-red, the shanks and feet bright yellow. Neither amateur or professional breeder will experience much difficulty in procuring from even moderate specimens the above requirements. At the same time, white frequently appears in the ear-lobes, and even in good exhibition stock. However, white in ear-lobes is not white lobes, as is seen in the Meditteranean breeds, but rather white in a more or less degree on an otherwise red ear. An absolute white ear, if ever seen, would be a disqualification when being judged; but if white in ear a deduction of from one to six points would be mentally taken off the bird's comparative degree of perfection, according to the quantity of white which appears. For the purpose of the standard all the sections, as has been shown, total 100 points, and many of the best breeders still think that as this white in ear-lobes is such an eyesore, ten, rather than six points should be lost for the defect. During the past few years, several Wyandottes have come out from England showing this objectionable trait, still they were excellent otherwise, and, perhaps, had it not been for this defect they would have remained in the more profitable show-ground where they were raised. This defect can be eradicated, for if in the male bird, and all the stock hens sound in ear, the defect will appear in but a few of the progeny, and if the faulty ones of these are rigorously excluded from the breeding-pens for two or three seasons it will rarely appear.

Should the original male bird have some excellent qualities, which the breeder desires to retain, he can be re-mated to the sound-bred stock, and although related the white ears will but seldom reappear.

The heads of both sexes have to be white, and it is very rare when any deduction has to be made here, the gravest case only meriting a five-point deduction. The neck in both sexes has to be silvery white, with a clear black stripe down the centre of each feather and free from tick. This latter term is not the insect which a neighbouring State imagines all the Sydney fowls are infested, and demands a clear certificate from officers in the Agricultural Department before such can enter the State in question. The tick of the standard, however, is to the breeder of choice Wyandottes more trouble than the prohibited article, many and almost perfect birds being spoiled from prize-taking by the dark spots or ticks, which often appear on the ends of the neck hackle feathers, and look as a dark fringe or border falling over on the birds' shoulders. This defect is in some specimens most pronounced, and is then called sootiness, and up to eight points can be deducted for this faulty colouring. The illustration of the pen of Silvers, facing page 346, will show the correct marking of this section of the exhibition laced Wyandotte. It may here be added that the white edging of the neck and hackle feathers should be pure white, and the black centre or stripe a dense black; a dull brown is a frequent occurrence and is objectionable. The back of the cock has to be pure white, free from yellow or straw colour, and the shoulder tip or butt white laced with black. The wing coverts well laced, forming two well-defined bars. On the breast and under parts the web white, with well-defined black lacing, free from double or white outer lacing, under colour dark slate, the true tail feathers and sickles black, thighs and fluff black or dark slate. It will thus be seen that the birds have to be black and white, but it is of frequent occurrence that in many otherwise good specimens this white is of a straw colour. This may be the result of exposure to the sun, but there are strains of birds in which this straw colour is implanted and it is most difficult to breed out, and those who are in possession of such, rather than attempt to get rid of it, will do better by disposing of every bird containing such blood and commence again with the now plentiful enough white and black strains. The saddle hackles of the cock should be the same as the neck hackles, this and the preceding constituting what is known in the standard as the back, and for which the faultiness in colour may be so pronounced that up to fourteen points can be deducted ; at the same time any bird suffering to such an extent, no matter how good otherwise, would scarcely receive a prize in any show. The under parts must have the feathers white with a narrow well-defined black edging, and free from an outer fringe of white, the lacing to be regular from the throat to the back of the thighs. In the endeavour to get this narrow lacing another evil has resulted, i.e., the lower part of the feather is right, but the upper part frequently comes without any lacing whatever.

The importance of a well-laced breast is evidenced from the fact that up to fourteen points can be deducted for this fault. The colour of the wings is also an important feature in exhibition specimens. The illustration shows two clearly-defined bars of black across the wings, frequently birds that are too dark in lacing for exhibition purposes excel in these wing qualities. Other conditions are to the colour of the flights and secondaries, but these rarely give trouble. The small soft feathers on the fowls thighs and stern are termed fluff. These should be black or dark gray, often they run to a silver-gray colour, and when so up to six points may be lost. The same number can be deducted for pale legs, while absolutely white legs are a disqualification. For white feathers in tail, or a bad carriage of such, seven points is the extreme penalty. For want of size and condition fourteen points may be deducted. Coming to the hen, it will be seen that owing to the difference in marking from the male bird the points allotted are slightly different in the various sections, but are as approximately alike as the framers of the standard could conceive.

The double-laced feathers referred to above, and which are a serious handicap, is a white or frosty edging, on an otherwise well laced feather, and most frequently appears on the breast of both sexes. Another fault in marking is the spangled feather, this taking the form of that seen on the Spangled Hamburgs. It is a heavy black marking on the feather, but does not extend all the way round. This fault is more frequently seen in the male bird. The mossy or peppery feather is the most serious of the mis-colours in the laced Wyandotte ; this is a small tick or fine black spot which appears in great numbers over the white centre of the feathers; if feathers of this sort are plentiful on either sex it is a handicap so serious that the birds need not be exhibited.

All the remarks and definitions referred to above in the Silver Wyandottes apply equally to the Golds.

In later-day breeding operations those desirous of securing the greatest number of show specimens in the progeny resort to what is known as the double-mating system, i.e., a cock or cockerel is mated with certain marked hens or pullets of a known descent, the object being to secure that the bulk of the cockerel progeny will be of well-defined markings, while the pullets may, to all intents and purposes, be valueless as exhibition birds, but most valuable to again mate for breeding good cockerels. The same process is employed in another pen where the object is for the production of exhibition pullets, the cockerels being only valuable to again reproduce similar results.

However, this scientific mating is outside the region of those for whom these papers are intended. The farmer who has secured a good pen of any colour Wyandottes need not resort to the above system, for with the standard already given, a brief experience will soon enable him to mate his birds in such a way as to fairly well produce their like, and by continuous mating and breeding, even from the same family, he can not only keep his stock up to the original colour, but retain all the vigour as well. The most sturdy and well-marked cockerels should be retained each year for breeding, and these mated to a pen each of the highest and darkest coloured hens, and the progeny of each pen duly noted will soon enable the breeder to secure results equal for his purpose to the scientifically-mated stock of the most successful fancier.

Chapter XI.

White Wyandottes.

White Wyandottes, without the slightest reservation, commend themselves as a farmer's fowl. The laced varieties, the Partridge, and other fancy colours may perhaps equal them in egg-production and

Mr. W. W. Smith's White Wyandotte Cock.
First and Champion at Royal Agricultural Show, 1905.

meat qualities, but, if not carefully looked after, will in a few seasons run out in colour and become of a nondescript appearance, a handicap that to the Whites does not apply.

The first thing to secure is type or shape; not the too common leggy sort, but those of full body, broad, deep breast, rather short shanks, and general cobbiness throughout. After this comes the colour, for just as there is a blue and a better blue, so there is a white and a better white. There are strains of White Wyandottes which are of a straw-yellow colour, and from these the fancier or other breeder need never expect to produce the much admired snow-white specimens.

The following is from the pen of one of the oldest English Wyandotte breeders, the Rev. J. Crombleholme, who writes on this variety: "If I desired to keep Wyandottes for utility purposes only, I should select the White. The White, as a rule, is the plumpest Wyandotte grown, for as there are no markings to breed for, but purity of white only, one need not fear to regularly introduce new blood in the year. As a consequence, the enervation of constitution that follows too much in-breeding does not exist and strong progeny is ensured. Another consequence of this freedom of choice is that the Whites are the best layers. Sweeping assertions of this nature are, perhaps, open to contradiction; but, at all events, my own best shaped Wyandottes are the Whites; they are also my best layers, and produce the most fertile eggs. In breeding Whites we must insist on purity of colour. It is no use trying to get good chickens from sappy parents. There is something in a 'sappy' feather which no one that I know of has been able to diagnose, and which is always perpetuated in young stock. Anyone, then, anxious to breed exhibition chickens must insist on a true white colour in the parents. No matter how big or how fine a cock or hen looks, if they are yellowish or discoloured keep them out of the breeding-pen. When I first began breeding White Wyandottes, I wrote to a noted breeder of White Leghorns, and asked him how he managed to show such *extremely* white birds, hinting that if there was anything in it he might let me know. His answer was that his was a *white strain*. I took it then that he did not wish to tell me his secrets, and let the matter drop; but now, after eight years of breeding, I have come to the conclusion that this white breeder was not joking, but telling a straightforward tale." Mr. Lewis Wright, in the last edition of his well-known poultry-book, writes on the White Wyandotte as follows :—" This variety is one of the two most generally kept in the whole of the United States, disputing with the Barred Rock alone the premier position in the American poultry world. This can only be the case with a white fowl where the poultry interest is chiefly in the hands of the farming class as it is there, and in that country the yellow leg is an added recommendation. But we can add our testimony that a White Wyandotte, besides being a most prolific layer, plucks to a most attractive-looking market fowl, and is most delicious eating."

In breeding Buff Wyandottes the same principle applies as in all other Buff fowls, the yellow shanks and beaks in these making the colour easier to get than in the Buff Orpingtons. A pen of even-coloured birds in both sexes will produce a large percentage of such. So far this variety is almost a stranger on the farms in this country; neither have the fanciers taken to it seriously.

Partridge Wyandottes, although of comparative recent development, are becoming plentiful, but so far are largely in the hands of the fancier, and whether they will be taken up by the farmer is yet a moot question. As egg-producers, like all the Wyandotte family, they are good, one breeder in this State positively asserting that a pen of four laid over 200 eggs each per year. This is more than double the number produced individually by a pen of this variety at the last College competition, but unfortunately for the variety there was only one pen competing, and just as in the other colours there are good and bad layers, and this particular pen chanced to be the latter. Concerning the merits of the Partridge as table fowls all the Wyandottes are good in that respect, and the specimens shown in last month's *Gazette* may be taken as a criterion of this breed of fowls as meat-producers.

The Silver Pencilled Columbian, and other new varieties like the Partridge, have not yet reached the farmer, and when they do possibly they will be found neither better nor worse than the earlier creation of this plentiful, pretty, and profitable breed of fowls.

Before closing it will be but justice to the breed to reproduce an article from the well-known American poultry judge and journalist, Mr. T. F. McGrew, contributed a short time ago to the *Country Gentleman* :—

"Just twenty-three years ago the original Wyandotte was admitted to the standard. This variety is known to-day as the Silver Wyandotte. A more descriptive title would he "Silver-laced" Wyandottes ; and of the Goldens, " Golden-laced." These distinctions are quite valuable in the way of description to the beginner or casual observer. When the Silver Wyandottes came into notice, they, like the Goldens, were managed not to make high quality exhibition fowls, but to make money by selling them broadcast over the land. If these two had been left to stand alone, they would have gone quite out of sight, like the Javas ; but, fortunately, from the original came the white specimens that, as White Wyandottes, have carried the whole family to fame.

" No family or breed of fowls possesses more real sterling worth than do the Wyandottes. This value is very prominent with the Silvers, and with the Goldens too. From the very first they have had value as egg-producers, as market poultry, for broilers, and as general-purpose fowls. No fowl excels them as a family in all these features of excellence These were quite as strong in the original Silvers and Goldens as they are at this time in any of the others. These two originals might have been quite as popular if it was not that to have real beauty of plumage is most difficult in their handling, and when clothed in poor or indifferent markings they are not pleasing to the eye.

" No one well informed as to their value would say that they are lacking in any of the utility features that are so valued in the Whites. This being true, one is rather forced to admit the value of plumage colour in market or utility poultry. How foolish any one is to write against the value of attractiveness in fowls ! If plain, unattractive appearance should guide, why not place equal value on the plainest-coloured Silver that is laid to the credit of the White? But with such overwhelming evidence of a general preference for the White, there is little chance to get away from this, and all kinds of reasons for placing to the credit of the White the greatest popularity of its family.

" As a favourite colour for the exhibition, for the true fancier and the novice, nothing excels Buff. When the Buff Wyandottes came into public favour, they held for a time the balance of power. They still command considerable attention in the show-room and with fanciers, but this has not and will not extend into what might be called general public favour. Buff Wyandottes will continue to grow in favour. They will extend into many new places; but, with all this, they will not supersede the Whites. Having equal value from a domestic point of view, the Whites have no advantage over them in this way. There must be other reasons for their not being on the crest of public popularity.

" We in this country favour the yellow-skinned and meated poultry. This we find to perfection in the Buff varieties. As to the question of plumage and pin-feathers, the Whites have no advantage over them, and, with all this, public favour follows the Whites. Since the coming of the Buff fowls into notice, no colour has been so attractive from the fancier's standpoint as the buff. The most talented have spent years at producing fowls of the true shade of colour. Page after page has been written on how to produce true buff, yet, with all this, their popularity as general favourites throughout the land has been confined to a limit within a circle of those who select them as an individual preference.

" The Black Wyandotte should be the prime favourite of all black fowls, but even they, with all their true worth, are seldom seen outside of the Boston and New York shows. They are as beautiful in shape, colour, and plumage as any of the Wyandotte family. When dressed, they are fully as attractive and delicate table poultry as are the Whites. They have equal value in every way, but they have not gained in public favour, nor is there much likelihood that that will occur. All they lack is public favour, for they have all the desirable qualities.

" The Partridge Wyandotte has the beauty of plumage of the black-red fowls. Clothed as they are like the Partridge Cochin, nothing could be of truer colour or more attractive character. They came among us with the blowing of trumpets and coloured illustrations; their club has done wonders for them; they are fully the equal of their kind; and while they are but a new variety, they have shown their attractive value, and have gained public attention; but so far they have not obtained the position of prime favourite, that is still held by the Whites, nor is there any evidence that they will.

" Following these are the Silver-pencilled Wyandottes, coloured and marked like the Dark Brahmas. No fowl of any breed or variety excel them in real domestic value. They are most prolific egg-producers, most attractive as show fowls. With all this in their favour, we cannot see why they should not push strongly to the front. At the same time we do not expect to see them supplant the White variety. If this might be, it will surprise all who have in mind the progress, past and present, made by the several breeds and varieties of fowls.

"Following these is the Columbian variety—a true Wyandotte that has the colour and markings of the Light Brahma. For them there is a future They will become the centre of attraction for a time at least. It must rest with the general public what position they will fill; but with them, as with others, we cannot see why they should supplant the Whites. At the same time, there will be a continued and increased demand for all these valuable varieties, and those who created and improved these many varieties of this valuable breed did great service toward better meat production.

"The Buff Plymouth Rocks are the single-comb rivals of the Buff Wyandottes. These two have gone along side by side, each having their following and both filling about equal position in public favour. It is recorded that the highest prices paid in late years for fowls have been paid for Buff Plymouth Rocks. They stand well to the front as exhibition fowls. None excel them in this, and they have gained this position as the result of careful breeding. They have every quality that belongs to any of the Plymouth Rock family. They have a right to demand at the hands of the general public equal recognition with any variety. But with them in the same family are the White and Barred varieties, which contend for public favour.

"Personally, we should prefer the White Plymouth Rock to all other American varieties. They have every advantage that belongs to the breed. One might think that the White Plymouth Rock would have as strong a hold on public favour as the White Wyandotte has, but this is not the case ; and while the White Plymouth Rock is highly valued, it has not gained equal favour with its rival of the Wyandotte family."

As previously said, the Silver-pencilled Wyandottes are the latest of the new variety, and not much literature is yet available on them. However, since my reference in an earlier *Gazette* appeared, the following has been contributed to an English journal by the oldest breeder of the Silver-pencilled in that country, Mr. O. S. Marshall. The illustration of these, which has already appeared, will show that this later sort for appearance is as handsome as any of the Wyandotte family, while for utility purposes Mr. Marshall's testimony should be conclusive.

"We sometimes hear murmurs, and read of others, decrying the addition to the family of Wyandottes. Some people do not hesitate to call them mongrels, and such would have us be satisfied with what we have already. They argue, the family is already large enough, with Silvers, Golds, Whites, Buffs, Partridges, Buff, and Blue laces, why not be content to perfect these? Happily, there are few such croakers, and it will be a bad day for the Fancy should their suggestions be seriously listened to. In the world, new discoveries and inventions are being continually met with. Some are good, and some bad, but they right themselves by the general law of the survival of the fittest. And so it is with Poultry. New varieties are being evolved by the fancier's skill, some of which are of the meteoric type, make a sensation for a season or two, and then sink into oblivion. I have in my mind a certain variety introduced from the Continent a few seasons ago, from which great things were expected, by some fanciers—I withhold its name for fear of hurting their feelings. Others continue to increase in favour, and make lasting friends wherever they are known. In this last division I place the Silver-pencilled Wyandotte.

"I do not claim to have had experience of every variety of poultry, but during over 20 years' experience I have never kept any to equal it as an egg producer. I regret now that I did not keep a strict account of the eggs laid by the four pullets in my best breeding pen during the last autumn and winter. They were hatched about the middle of March, 1904, and commenced to lay in September, and they continued to lay all through the winter, without becoming broody until the beginning of April, when one wanted to sit, but was easily checked, and, after a few days, resumed laying. Keeping two other varieties—Partridge and Columbian Wyandottes—I can compare them as layers, and while the two first-named were either stopped by the severe weather, or resting, the Pencils supplied me regularly with their tinted eggs, when the market price was six for the shilling.

"So much for the utility side (and, to my mind, no breed will long hold the public favour without it has utility properties), but read what the exhibition tale is. One pullet was 3rd Dairy, 4th Newbury and Grand International (Alexandria Palace), 3rd York, Bromsgrove, and Banbury. Her sister was 1st Scottish Wyandotte Club, Edinburgh, 1st and special Banbury (here she was given special over the cock class as well as her own sex); the other two pullets were unshown. I mention this because some people have an idea that exhibition birds are not good layers. To such people, I say, try Pencils and you will change your opinion.

"The shortest description of what they are like would be to describe them as having the exact colour of the dark Brahma, with the general shape, size, and characteristics of the Wyandotte. Like the Brahma and many other laced varieties, to breed exhibition birds it is absolutely necessary to resort to double mating. By this is meant the selection of a pen of birds to breed good males, and another pen to breed good females. Briefly, for breeding good males, the lord of the harem must be as near the exhibition type as possible, while his mates need not have the characteristic pencilling and colour of the exhibition females, but they should have good combs, well striped neck hackles, yellow legs, and be as large as possible. For breeding females, the hens or pullets must approach the standard type of the exhibition females. The lord and master of this pen must be pullet bred. That is, he must have been hatched from a similar mating, and if his mother happens to have been a winner, so much the better. He will be of no use as a show bird, because the standard requires him to have a black breast, while the experienced breeder will say, "Give me one for my breeding-pen with white tickings or lacings on his breast and hocks.

"Some people have tried to argue that this double mating is against the popularity of breeds requiring it, and that it ought to be discouraged; but those who wish to have birds fit to win must do it. Time is everything now, and of what use would it be trying to exploit a theory or fad, while our neighbour, breeding the other way, takes all the prizes? One can easily cite as an example that this double mating does not stop a breed's progress, by calling attention to one of the most popular varieties of the day—the Partridge Wyandotte. Nobody but a novice would expect to breed exhibition birds of both sexes from one pen. I am fully aware that occasionally an exhibition cockerel comes from a pullet-bred pen, and I now have such an one in my yard, but it is one of those freaks of nature one sometimes hears of. The double mating here has not stopped its popularity, as each year it advances by leaps and bounds. It is in the recollection of your readers that only last year at the Grand International Show, a plucky lady—Miss Rilot—gave the hitherto unheard-of price of £165 for a soft-feathered cockerel, and during the season others changed hands for over £50. I admit it would save a lot of trouble if it could be done without resorting to two pens, but if it cannot—well, it is Hobson's choice. The idea is not a new one, as breeders of the Brahma, &c., used it years ago.

"As in the case of the other branches of the Wyandotte family, the Silver Pencil was produced in America and imported into England. To Mr. Brackenbury belongs the credit, I believe, of their origin, and there is no secret how they were produced—mating Wyandotte cocks with dark Brahma, lightly-feathered hens. I believe I am right in saying that this method is a quicker way of attaining the desired end than the reverse mating. To that well-known Yorkshire breeder of Wyandottes (Mr. John Wharton) belongs the credit of first importing them early in 1901. At a later period, Mrs. Campbell and Mr. J. Pettipher founded a strain of their own by a similar mating. The American birds were beautifully pencilled and striking in appearance, but they were undersized, and breeders found it very necessary to introduce Brahma blood to increase the size. This has been judiciously done with gratifying results, producing male specimens weighing up to 9 lb. in weight, a most satisfactory result. Remembering the large number of entries at our leading shows last season, it is remarkable the way they have won favour generally in so short a time. With their splendid qualities to recommend them, one does not require to pose as a prophet to predict still greater success for them.

"At the beginning of this year we thought it was quite time a standard for the variety was drawn up, and also that a club be formed to look after its interests. With that end in view, a meeting was called and held at the Banbury Show last January. The standard has lately been issued, and all the principal breeders have joined the club. The members now number forty, which speaks well for the future.

"As I mentioned earlier, the colour follows that of the dark Brahma, with the bands, to be as numerous as possible, following the shape of the feathers, on a steel grey ground, in the female, while the shape follows the accepted type of the Wyandotte. As regards the weight of the birds, it is rather curious that it has been left to the club to put them up; that is, we think that the Wyandottes generally are underweighted. The Pencil standard reads, ' Size and weight, rather large, matured cockerels about 8 lb., adult cocks about 10 lb., pullets, 6½ lb., adult hens, 8 lb.' Comparing these weights with the other standards, they will be found to be from 1 lb. to 1½ lb. heavier, and already some fanciers are up in arms regarding it. I would ask them to remember that it has been somewhat of a taunt, generally, against Wyandottes, that they were very nice and attractive birds, but rather small, and now that breeders have increased the size, the standard must be put up, otherwise a judge ought, as a matter of fact, to penalise birds over the standard. Anyone frequenting the shows will readily admit that about all the prize winners of late in Whites and Partridges would have to be put back if this rule had been literally interpreted. Rules must be made for what birds have to be, and not what they are at the time. As our rule was made by prominent Wyandotte breeders, we quite expect to see the rules for the other varieties altered and brought into line with us.

"In conclusion, let me strongly recommend anyone on the look out for a change in their birds to give the Silver-pencil a trial. I am certain it will answer every reasonable expectation, and give them a lot of pleasure and profit."

Pages more could be given of the high opinions held by American and other breeders on the merits of the Wyandotte family of fowls, but sufficient has already been shown, the writer's opinion and experience of them being briefly as follows :—Wyandottes are in every sense of the word a farmer's fowl. There are other breeds and varieties essentially as good, and those who breed other varieties and do well with them should continue doing so. There are, however, some who do badly with other breeds, and if such purpose making a change, then they could do very much worse than give the handsome easily confined Wyandotte a trial, and should their efforts with the new breed still be unremunerative, the privilege to them remains of falling back on the well-worn aphorism, "It won't pay."

CHAPTER XII.

ORPINGTONS.

In 1899 I contributed to the *Gazette* a series of lengthy articles on the above then comparatively new breed of fowls, and having been a close observer of the breed's intrinsic worth from the date of its introduction to Australia, about a dozen years previous, I had then but little hesitancy in supplementing the nomenclature with a prefix, the proved worthiness of the breed since that time justifying the adjective so much that at the present day this breed of fowls are most generally written or spoken of as " The Popular Orpingtons." The title is now well-worn, but none the less justified; indeed, the diversity of the breed's patrons, and its universality amongst those whose conditions are most favourable for profitable poultry keeping, that was there any desire to descriptively enhance the value of the breed, then such might be by way of the affix, "A farmer's fowl."

It should be here mentioned that there is no desire on the part of the writer to unduly inflate the commercial worth of any breed, the sole object being to bring into prominence the salient points of the different varieties of fowls, and to thus enable intending purchasers to discriminate between the claims of such, with a view to obtaining the sorts best suited for their respective conditions or requirements; and when speaking or writing of fowls, I have always studiously

Mr. L. L. Ramsay's Black Orpington Hen.
Winner at Royal Agricultural, 1905, and other Shows.

avoided emphasising any as the best and only breed; for so sure as such a thing has been done by any writer, the first practical test usually upset the contentions. So far as laying properties are concerned, assertions have been made that a certain breed or breeds are the best. However, did such require refutation, the various laying competitions in the several States have conclusively shown the weakness of such statements; for not only have different breeds made the best performances at almost every test, but at the same test that such breed won, other pens of the same breed frequently occupied the lowest position. Speaking generally, there are a few well-known breeds better layers than some others; but then these better ones when put to a test have been disappointing, and have on various occasions laid actually fewer eggs than did what are called the poor or moderate laying breeds. The whole thing is this: Orpingtons, Wyandottes, and the Mediterranean breed can be relied on as the best layers we have; yet the farmer or other party desiring to commence poultry keeping for profit must not think it is sufficient to secure a setting or two of eggs or a pen of any of these breeds, but rather a good laying family or strain of such; nor is this the result of any lessons from the competitions in the several States, but has been the experience of all practical breeders. Not only that, but hundreds of years ago the then writers on this subject realised that there were good and bad layers among the flocks. In a " Treatise on Husbandry," written by "Maystre Groshede, sometyme Byshop of Lincoln," compiled by the learned bishop for the guidance of Margaret, Countess of Lincoln, who was left a widow A.D. 1240, it appears instructions were to be given to the farm bailiff. From the wording, it is conclusive as to the yield and profit that was expected from well-selected poultry. " Fyve hennes will bring in IIIs. (three shillings) in a yere. I shall prene it by reason, for in halfe a yere be XXVI weeks, that is IX score dayes, and in eche of these dayes shall have an egg of eche henne in that half yere, and XXX eggs be worth a penney. He concludes : " Every henne shall answere you of IX score eggs or of chickens to ye value." Many other early writers frequently refer to good laying hens, one remarking that on a Kent farm in 1790 a hen was known to lay 230 eggs. Coming down to Moubray, the latest edition of whose work was published in 1854, there being then several well-known breeds of fowls, writing of Hamburgs he says : "They occupy the first rank as layers, the careful trial of a pen of them showing 230 eggs, and another pen 256 eggs each hen for the year." Many instances are recorded by modern writers of extraordinary laying of individual hens, and all emphasising that the feature of laying is governed other ways than by the names the fowls are known by, all unprejudiced writers being conclusive that as egg-producers there is actually no best breed. The same remark applies when a fast-growing carcase is desired. Several breeds have claim to this, while there are people who advocate crosses; whereas the actual facts are that, when tests have been made, the bulk of the claims have been dispelled and the theories exploded, all showing that strain, management, climatic, and other conditions are the factors which contribute to best results.

In the article on the Wyandotte fowls, I have said that, whether for egg or meat, that breed is most profitable to keep, the egg illustrations showing the actual performance and profits, while the photographs of the two dead fowls, reared under my own supervision, proves that the breed, for whatever purpose, is at least one of the best.

The one which I now purpose dealing with is also an excellent fowl, and although a shorter time before the public, is bred by fanciers and utility men alike—by suburban householders, orchardists, poultry and other farmers, and has penetrated to the very uttermost homesteads of the State; indeed, so widespread has the Orpington become in the comparatively brief period it has been before the public, that there is no hesitancy in placing it in the front row of farmers' fowls.

CHAPTER XIII.

Briefly Historical.

As in many other things and matters, the latter half of the present century has witnessed a complete revolution in not only poultry-breeding and management, but also a multiplicity of breeds far in advance of any other kind of stock; indeed, to those not immediately connected with the poultry fancy, the almost yearly announcement of new breeds and new varieties of old breeds prompts the query—What are the circumstances which warrant this annual innovation? the list being now of such inordinate length as to embrace about 100 breeds or varieties of domestic fowls.

The mysterious part of this rapidly-increasing catalogue is the fact that each breed is heralded on its introduction as something in every way superior to its predecessors, and, without questioning the statement that each new production embodies some useful quality lacking in hitherto established breeds, we cannot get over the fact that, did all the new breeds justify, and continue to justify, the characters which accompanied them on their arrival, the latest manufactures in the poultry world would be capable of laying several eggs a day all the year round. Indeed, for the past sixty years there has been a regular procession of new breeds of fowls, each one, according to those responsible for its introduction, being destined to fill the long-felt want, namely, a bird of fair size, with a rugged constitution, a quick grower of good white flesh, small feeder, and a layer of great quantities of eggs. Whether this want has yet been filled is still a moot question; and even did any of the present day breeds meet the above exactions, there is not a doubt but that in the near future other breeds or varieties would be produced with even greater claims for public recognition and favour.

Whatever difference of opinion exists amongst naturalists and others, the many who have written on poultry, as to their origin, &c., all appear to have agreed that it is impossible to fix the period of their first domestication. It is generally considered as correct that the first notice we have of the domestication of poultry is when we read of King Solomon consuming daily in his household "fatted fowls," while several hundred years after the Prophet Nehemiah had prepared daily for his household an ox, six choice sheep, and "fowls." Many later allusions to fowls are recorded by the ancients. Pythagoras (B.C. 535) refers to them as used for sacrificial purposes. Cicero (B.C. 106), too, informs us that hens were reared and kept entirely for the profit derived from the sale of their eggs; while, according to Pliny, the inhabitants of the Isle of Delos were the first to fatten their fowls by artificial means (cramming, we may suppose), and it was from them, he adds, that the rage for devouring fowls loaded with fat spread like a contagion amongst the luxurious gourmonds of his time, who spent their lives in endeavouring to produce some unheard-of dishes.

These references, however, are only to fowls—what shape, type, colour, or conformation, we know not; but that they were amenable to the fattening process is evidence that, even at that remote period, the poultry were capable of being fattened, which is more than can be said of many at the present day.

At what period domestic poultry became to be differentiated into breeds or classes and named, is involved in obscurity. Certain, however, it is that cock-fighting was one of the sports of Ancient Greece, and several Grecian cities were celebrated for the fighting qualities of their fowls. The sport was also adopted by the Romans about 450 years before the Christian Era, or, as some authors tell us, soon after the Peloponnesian war. One eminent historian tells us that about this time they had a breed of hens at Alexandria, in Egypt, celebrated for their fighting qualities. And even assuming that these fighting-birds of the Greeks were the common or domesticated fowl, the battles were invested with such interest, by reason of the political or religious issues involved, that we may expect the victorious cocks would be specially selected for producing others of like valour, and this process being continued—of breeding from the victors only—would soon produce birds of superior fighting qualities, courage, and endurance; and assuming that such was the case, this family, bred for fighting properties, can, I think, be safely called game, and consequently the first breed of fowls we know of, apart, of course, from the domestic, from which they were bred.

The introduction of game fowls—*i.e.* cock-fighting—into England has been ascribed to Julius Cæsar, but the earliest distinct notice is first found in a description of London by William Fitz-Stephen, a writer of the time of Henry II, who states that the pastime was so generally in vogue that it was the customary game even of school-boys. In the reign of Henry III the sport had become so general that it was found necessary, in 1366, to check it by proclamation. Cock-fighting, however, continued to flourish, royalty itself, in the person of Henry VIII, being one of its patrons, a cock-pit being added to Whitehall by that monarch for his own amusement.

In consequence of the mischievous results attending cock-fighting, Queen Elizabeth, in 1569, was obliged to issue a royal proclamation for its suppression, but with little better results than that attending the ukase of Henry III, 200 years before. It is even recorded that the rigid Scotchman, James I, indulged himself twice a week in the diversion of the cock-pit. And although legally suppressed, the sport has been clandestinely carried on in some of the rural districts (England) to the present day. The sport was introduced to this State in the very early years of its history, the old records stating that in 1798 the Sundays about Parramatta were principally spent by the majority of the inhabitants in cock-fighting ; and it is worthy of note that at the present day the district from that to Windsor and Hawkesbury is celebrated for the superiority of the poultry which it sends to the Sydney market, "Hawkesbury Chickens" always realising shillings a pair more than those from any other locality. Our present Colonial game evolved from the old fighting stock of the early days, 80 or 90 per cent. of the fowls bred on the Hawkesbury being of this breed.

The above brief historical facts show that for centuries game (fighting) fowls have been in existence in England, and are consequently the first breed of which we have any record, and whether they evolved from the earlier domestic fowl, or a result from a cross with some of the wild jungle fowl, matters not; they head the long procession of breeds which have since followed, one of the latest being that which gives the title to this paper.

Dorkings are also a breed of great antiquity, one writer at least claiming them as existing prior to the game. At any rate, when Shakespeare makes Justice Shallow, of Glos'ter, order a couple of short-legged hens for his guests' repast, it is considered Dorkings were alluded to. When they were first called Dorkings is involved in obscurity, but can be traced back for over 200 years. However, sweeping all conjectures aside as to the date of origination, the fact remains that at the beginning of the present century we had two acknowledged breeds of fowls, and both at that time noted for qualities which to the present they still possess—the old English Game, a fighting breed, and the English Dorking, noted for its good eating qualities.

A few years afterwards, about 1806, an old agricultural work mentions one or two additional varieties. Following this comes what is frequently described as the first work on fowls, entitled "Moubray's Treatise on Domestic Poultry," first published in 1814. This now rare work, showing that in the few years prior to that date, although poultry shows were unknown, there were poultry-breeders experimenting with the then few known varieties, to which they applied local names ; however, these new creations quickly disappeared, and were heard of no more. The following twenty years were fertile in new breeds, imported or otherwise, those then known being Game, Dorking, Chittagong or Malays, Polands, Bantams, while a black fowl was occasionally seen in several English counties, which is stated to be the ancestor of our present Minorcas. The latter ten years of the first half of this century was the period of all others noted for new breeds and varieties, warranting the author, whom I have already quoted, issuing an enlarged edition with coloured plates of the then best known breeds, namely—Polish, Spanish, Malays, Scotch Bakies, Andalusians, Turkeys, Guinea Fowls, Aylesbury and Muscovy Ducks. Following these came what is known as the Asiatics, *i.e.* Brahmas, Cochins, &c., these varieties shortly after their arrival taking

their place as head of the then poultry list for their commercial qualities, the poultry authors of that time, Doyle, Nolan, Tegetmeier, &c.,
all writing most approvingly of the breeds.

CHAPTER XIV.

Excellencies and Deterioration of the Old Breeds.

MR. TEGETMEIER, writing about 1852, says : " The Brahmas have
the reputation in the States of far surpassing the Cochins as layers,
and of being most excellent mothers. The chickens are remarkably
hardy, of rapid growth, and feather quickly, and in plumage and
colours most closely resemble the parent bird, a circumstance which
goes very far to prove that they are a distinct variety to the Cochins,
as the fact that the latter cannot be bred true to colour is well known."

Mr. Ferguson, in his cleverly-written work on " Prize Poultry," of
about the same date, says: " The flesh of the Brahmapootras is
superior in quality to that of the average Shanghais or Malays, is of
good flavour, white, plump, and juicy, with less offal, and having the
advantage of superior weight over the Dorking ; while their eggs are
larger than those of the former birds, and more abundant than the
latter."

In 1853 there were three pens of these exhibited at the Metropolitan
Show, London, and on these the editor of the *London News* remarked :
" There is a class of fowls which seem likely to outrival even the
Cochin China themselves. They are the Brahmapootra fowls. Not
only with regard to the superior quality of flesh, but from the quantity
of meat they have on the breast, they are considered to be superior to
the Cochin China. The average natural weight per pair is said to be
from 22 to 25 lb., the cock ranging from 11 to 15 lb."

Nor can it be said that these useful qualities were connected with the
breed only on its introduction, for fifteen years later, in the 5th edition
of Mr. Lewis Wright's popular work, it is stated : " With regard to the
economic methods of Brahmas, the pullets lay when six months old,
and usually lay from 30 to 40 eggs before they seek to hatch, but I
have repeatedly known pullets begin to lay in autumn, and never stop
—let it be hail, rain, snow, or storm—for a single day till next spring.
As to their size, I have had a cock weighing 15 lb. and hens 12 lb.,
but these are unusual weights. I have, however, two cockerels of this
year (1866) only six and a half months old, one of which weighs 10¾
lb. and the other 11¼ lb. I consider 12 to 13 lb. for a cock and 9 to
10 lb. for a hen very good weights. Cockerels for exhibition, when
six months old, ought to weigh from 8 to 8½ lb., and pullets from 6 to
7 lb." The above testimony was contributed by a noted breeder of
that time, and was confirmed by the author of the work in the
following extract :—" With regard to the merits of Brahmas, they
rank very high ; in size the dark variety surpasses every other breed
yet known. They lay every day in the depth of winter, and scarcely

ever sit till they have laid 30 or 40 eggs. As winter layers no breed
equals them. We are writing at the end of November, and have a
hen which has laid 45 eggs in 48 days, while others are little inferior.
Brahmas are likewise very hardy, and grow uncommonly fast, being,
therefore, very early ready for table, in which particular they are
profitable fowls, having plenty of breast meat."

Although grossly exaggerated stories were told of the productiveness
of the Cochin when first introduced, and a poultry mania in consequence
arose, when this mania subsided, and breeders were able to give a
calm and unbiassed opinion of the breed, it was testified to as a most
excellent one. Lewis Wright, in one of his earliest editions, says :
"The chickens, though they feather slowly, are hardier than any
other breed except Brahmas, and will thrive where others would
perish; they grow fast, and may be killed when twelve weeks old.
They do well in a confined space, and cannot fly over a 2-ft. fence.
As sitters and mothers they are unsurpassed. They are prolific layers,
especially in winter, when eggs are scarce and dear."

Spanish, another of the old breeds, were world-renowned as good
layers. Polish also laid abundantly, while the Hamburg was in its
earlier history known as the Dutch every-day layer.

Coming to those of more recent origination, the same tale has to be
told; the character which accompanied them on their arrival, or
introduction, was a most excellent one, and for some time their cre-
dentials were thoroughly justified. Some breeds were noted for one
good quality, other breeds were notorious in another direction, while
some had the reputed embodiments of everything which go to constitute a perfect fowl. That the majority of the breeds were possessed
of the various economic merits there is not a doubt—as the quoted
authorities certify, but that these merits are to a certain extent in
some breeds partially lost, and in others thoroughly so, is also true;
the cause of this retrogression and its bearing on the origination of
the Orpington I shall endeavour to show.

Of the multitude of books issued on behalf or in the interests of
poultry in England, the favourite apology for their appearance is the
huge annual bill of several million pounds paid by that country for
foreign eggs and poultry, the object of the works being to lessen this
bill in favour of the home producers. Some of the writers have
fallen into the error that these immense importations are of recent
growth. Such, however, is not the case, as the following extracts
will testify :—Mr. Legrand, a member of the Statistical Society,
as far back as 1813, tells us that in that year France exported to
England one and three quarters millions of eggs; in 1822, they had
risen to fifty-five millions, and in 1834 had increased to sixty millions.

A writer in *The Penny Magazine* in 1837 calculated the imports of
eggs to England for 1836 from all sources as sixty-nine millions, the
British revenue that year benefiting to the extent of over £24,000 by
the then 1d. a dozen duty, while Mr. Weld in his statistical survey
of Roscommon, stated that £500 was the daily sum paid by England
to Ireland for eggs alone. These importations to England kept

increasing until about 1845, and for that year France received from England £250,000 for eggs. The foregoing figures and extracts are to show that even at that early date the breeding and rearing of poultry was of much more importance than is generally supposed. It should also be stated that the French importations formed but a small part of the English egg consumption, for independent of her own supplies the Irish shipments to London and Liverpool in 1845 had reached nearly forty-eight millions, the value of which, at the average price of 5s. 6d. per 124, the then price, gives us a sum amounting to about £122,000. The amazing growth of these figures began to cause uneasiness in England, when most opportunely the Brahmas and Cochins came upon the scene, with laying and other qualities so wonderful that the foreign egg trade was considered by some enthusiasts as good as doomed. The introduction of these new breeds was the cause of a complete revolution in poultry-keeping ideas. ·

Poultry societies were established in most of the large English towns for the encouragement of the best breeds, &c.

Agricultural societies found it incumbent for them to make provision for poultry in their catalogues—indeed, the many merits which characterised these new breeds awoke a general interest in poultry-keeping, which has gone on increasing to the present day; but remarkable to relate, although the Cochin craze, the Brahma boom, and the multiplicity of shows were responsible for thousands of recruits to poultry-breeding, the importation of eggs and poultry to England, in place of being checked by the admitted productiveness of the above, and other new breeds, steadily increased, and why this has been so has often been told by the authors who first noted their many admirable qualities. Some of these same authors are still alive, and now lamenting the decadence of their earlier favourites, have no hesitation in giving reasons why these older varieties have deteriorated, and are obliged to take a back seat in favour of those whose commercial claims are more pronounced.

Chapter XV.

POULTRY SHOWS.

Some fourteen years ago in a prize poultry essay, when contrasting the English poultry-keeping with the French, I described the latter as a country of poultry-breeders and the English as a nation of poultry-fanciers, a slight distinction with a very great difference.

The French peasants breed their fowls for economic qualities, and net their many millions a year profit, several of these millions coming out of the pockets of the English people. The latter, generally speaking, do not trouble themselves about the common commercial side of poultry-breeding, but rather keep them as a hobby, as they do their dogs, pigeons, canaries, and other pets for the interest and amusement they afford, and for the honour and glory attached to winning the prizes, cups, medals, and other honours for which they at their great shows compete.

This competitive spirit has grown to such an extent in the old country that every town of importance has now its poultry show ; a late issue of one of the fancier's journals having over 100 of these exhibitions advertised to be held within one month, while during the past year over 700 poultry shows were held throughout the United Kingdom.

The first object of the majority of these fancier's institutions, as stated in their rules, is to encourage and develop the breeding of poultry, and that they do this is evidenced by the enormous support they receive in the way of entries. The valuable money and other prizes offered at these shows provoke a rivalry of the most lively character. It is at these exhibitions breeder meets breeder, when they talk about and study each other's productions, and where the various breeds can be seen, and the specimens examined as to which is superior in colour and shape of its variety, the perfection of this shape and colour being the embodiment of the fancier's art and the fulfilment of the first object of the society. These shows are the great festivals of the fanciers, every breed and variety is brought together, exhibited in spotless condition, compared with each other by the judges, become decorated with well-deserved prize cards and ribbons, &c. Thousands of visitors cheerfully pay an admission fee to witness these great gatherings of feathered stock, whose effect is that not a show passes but many recruits are added to swell the great army of poultry-breeders. Unfortunately, however, for themselves and the poultry industry generally, very many of these annual recruits imagine that the object to "encourage and develop" refers to that for all qualities, economics, &c., or, in other words, that the prize-winning birds being the best birds are the best birds for practical purposes ; such, however, is not the case, all other stock, as cattle, sheep, horses, &c., are awarded prizes for the possession of certain points, these points being an index to qualities for which the class of stock is noted ; not so with fowls, whose awards are made according to standards that have no relation to commercial qualities. The birds are judged for appearance only ; the knowledge of any cock in the show being sterile would be no handicap to him receiving the championship, while not infrequently the first-prize hen has long prior ceased to produce an egg.

Fancy Points *v.* Utility.

In a previous issue of the *Gazette*, I pointed out to poultry-breeders for profit that stock for such purposes need not possess show points, but rather the reverse, and quoted authorities to prove this, which, to amateur breeders or those lately entered upon the pursuit, was a surprise. No so with what may be called the genuine fanciers, who breed their birds for pleasure and admit it, although my then remarks on that subject received the general approval of the state, interstate, and foreign press, one writer challenged the statement on the grounds that my authorities were old-fashioned, behind the times, &c. However, the apparent strictures in no way prompts these allusions to the subject, but rather the present article on the Orpington would be far from complete were not an extended reference to the Fancy *v.* Utility subject be made, inasmuch as this, of all others, was the one thing which inspired the late William Cook to attempt making a fowl which should be not only a good one but *continue* to be a good one, independent of, in spite of, or with the assistance of, the fancy or fanciers.

In his work, "Fowls for the Times," the late Mr. W. Cook says :— " When I commenced poultry-keeping many years ago, a few breeders kept good birds. They were, however, bred just for type, and scarcely ever with due regard for utility, until the birds which won in the show-pens were—as indeed they have been, all too sadly, ever since— just the very worst of layers, and sometimes the worst of table-birds ;" and further, " One of the things we have yet to learn as a fancy is that it is the fowl, not its feathers, which form the chief value of the bird itself. I therefore set to work, and, by careful and judicious breeding, I was able in a few years to give to the poultry-breeding a breed which, for egg-production and table, has been accepted as the grandest products of the poultry-world."

The authorities I quoted in my previous articles were certainly old, but this was one reason why I selected them—a lifetime in the poultry world and through its many vicissitudes. Mr. Tegetmeier was in the heat of the fray when the Brahmas and Cochins arrived, and then wrote of their many excellencies. From that date, now over fifty years, till the present time, this authority has been actively connected with the poultry industry, poultry shows, poultry clubs, and other poultry institutions, and the poultry press, and has been a witness of all the work which tended to bring the above and other breeds from their one time eminence and high estate to be now the most neglected of all by those who keep poultry for profit.

As previously said, very many of the recruits to poultry-keeping brought about by their visits to the show vainly imagine that the legitimate aims of these shows is to improve the breeds in a profitable sense, the misconception being too often responsible for their short-lived enthusiasm. This misunderstanding occasionally crops up even in the old country, the editor of a leading fanciers' paper there some time ago being obliged to put an end to correspondence by the following :—" We now take up this subject, not so much with a view of adding anything to what has been said by either side, but because this discussion, being a prominent one, may be considered, to a certain extent, as typical. It illustrates the fact that the show-bird and the useful bird are two distinct things. As we have already said, we think there is a confusion of ideas, and the sooner this confusion is cleared away the better it will be both for fanciers and for those who breed poultry with a view to profit. A sort of tradition lingers round the show-pen that the great object of poultry shows is to improve the breeds of poultry from an economic standpoint. This, we think, is a mistake. The legitimate object of poultry shows is simply the encouragement of a most interesting pursuit, which may be followed either for pleasure or profit, according to the taste of those who follow it. Poultry shows, no more than pigeon shows, cage-bird shows, or rabbit shows, have anything to do with poultry from the farmer's point of view. It is true that fanciers themselves encourage the mistake to which we have referred, as it aids them in disposing of their surplus stock to those who do not keep poultry for exhibition, but merely for laying or table purposes. We do not think, however, that this deception is intentional. There is no doubt that poultry shows have done much to spread abroad through the country pure breeds of fowls; and as breed after breed noted for its laying qualities or table qualities in the places from which it has been introduced has come to the front, it has naturally acquired a popular character apart from its exhibition character, and this popular character has clung to it in many instances long after the original economic characteristics of the breed have been lost or impaired by breeding for fancy points. At the last Birmingham show a coloured Dorking hen was sold for £25. Will any one assert that this price was given, or would be given, with a view to its table qualities?"

The above editorial article, as can be supposed, conclusively settled the controversy, but not for all time, as it breaks out occasionally at irregular intervals. The discussion in almost every case being brought about by short-experience breeders, and always silenced by those who have been longer at the game frankly confessing that appearance only was the goal of their ambition.

However, this subject would not receive justice did I leave out two giants in the fancy poultry world, namely, Lewis Wright, and the late Alexander Comyns, LL.B. The former, who, by his great work on poultry, has done more for the fancy than any other author. Wright's poultry book is known wherever poultry shows exist, and is considered by many the standard work of reference on all matters affecting the poultry question. This work was written by a fancier for fanciers, still the author fearlessly embodies the following in his chapter on "Poultry as a National Food":—"For reasons we shall point out in next chapter, it happens that the fancier of poultry, in whose hands the cultivation of pure varieties has chiefly lain, has, for the most part, sought to develop other qualities than those which are of most importance to the commercial poultry-keeper. He seeks principally for feathers, and as his best birds in point of colour will seldom be also the best layers or fatteners, these points are comparatively neglected. It can hardly be doubted that from these causes some pure breeds, taken as a whole, have actually deteriorated in economic value of late years. That some Houdans and Brahmas, for instance, do not lay so well as these breeds formerly did; and the point to be specially kept in mind is that the commercial producer, by making his selections in the same way as the fancier, but with reference to other points, may attain the same success.

Chapter XVI.

ORPINGTONS.

Shows, &c.

In connection with the good qualities of the old breeds, and the admitted deterioration, through over-showing or other causes, the opinions of the late Mr. Comyns are worth reproducing. Mr. Comyns was an exhibitor, reporter, judge, lecturer, &c., and for several years Secretary of the English Poultry Club (was my own nominator for membership of that club in 1884, a coincidence being that the late Mr. W. Cook, the man who made the Orpington, was also nominated at the same meeting). In fact, Mr. Comyns occupied every honourable position which the fanciers' poultry world of late years offered, and was of all others competent to speak of its results. In 1886 Mr. Comyns was engaged to deliver a series of lectures on poultry-keeping, under the auspices of the Institute of Agriculture, at the Museum of Geology, London, and, though editor of a fanciers' paper, pluckily introduced the subject as follows :—

" Throughout France poultry are almost universally kept for one of two purposes, either to lay eggs, or to fatten for the table. In England, on the other hand, a very large proportion of the poultry are not kept for either of these purposes, but to gratify the taste or 'fancy' of their owners. It is here, I think, that one of the gravest causes of want of success in poultry-keeping in England is to be found. There are no statistics of poultry-keeping in England available, but I hardly think I am beyond the mark in saying that more than one-half of the poultry of this country are either themselves fancy birds or immediately descended from such. Some of my readers know perfectly well what I mean by fancy birds, but, as I am here to instruct those who do not know, I will explain my meaning.

" In this country there are some hundreds of poultry shows in the course of the year, where prizes are offered for the various breeds. At these shows the birds are judged according to certain standards, which have in the course of time come to be accepted as the ideal of perfection of each breed. You will understand this better later on, when I come to describe the points of various breeds, but I may say here that these points of perfection are, as a rule, arbitrary and useless in themselves, from a practical point of view, as could well be imagined. Feathers—laced, pencilled, spangled, barred, striped, &c.; combs—single, double, triple, rose, horned, and several other sorts; legs—feathered, bare of feather, yellow, black, blue, green, white.

" All these points have their value, and no slight value, in the estimation of the fancier. I have been a fancier, and have found it a most engrossing occupation or amusement; and I do not for a moment mean to say a word against fanciers as such. They have undoubtedly in some respects benefited the poultry-keepers in the country, but it is manifest that their pursuit in itself is of as little practical utility as would be the desire of a dairy farmer to produce cows with their horns twisted three times round, or some other similar arbitrary point not in itself an indication of purity of race. As an amusement, or as a pursuit from which money may be made, poultry-fancying is desirable and beneficial; but it is manifest that in the pursuit of all these fancy points the useful is sure to be lost sight of. What is necessary for the practical poultry-keeper is a breed of fowl which lays the largest number of eggs in the shortest possible time, and which will fatten readily and be of good quality on the table. The poultry-fancier, in the pursuit of his lacing, pencilling, &c., is obliged to make such a trivial matter as laying a very secondary consideration."

Mr. Tegetmeier, F.Z.S., editor of *The Field*, and author of "Profitable Poultry" and other works on this subject, and a judge of table poultry at most of the leading English shows, says :—"I have seen with regret the steadily increasing tendency of poultry shows to encourage mere fancy varieties, and to ignore altogether the profitable value of the birds exhibited. This has gone on to such an extent that I do not hesitate to affirm, as the result of my experience of half a century, that no one breed of fowls has been taken in hand by the fancier that has not been seriously depreciated as a useful variety of poultry. Further, what I object to is that fancy poultry should, in this country at least, take the place of useful birds that are fitted to supply the markets with poultry and eggs, for, as at present conducted, fancy points only have to be considered by the judges, the result being that the economical value of many breeds has been entirely lost. For example, Spanish, from being abundant producers of large white eggs, have become very indifferent layers, some of the notorious prize winners being sterile. Cochins on their introduction were good layers, and are now the worst. Dorkings, that formerly supplied the best fowls for London markets, are now bred as show-birds, and are not equal to Surreys. Game, formerly bred for table fowls, are now elongated out of all knowledge, and look more like the waders of the ornithologist. I wish to show that for economical purposes it is absolutely necessary that such views should be set forth, for at the present time our agricultural societies are doing what I conceive to be considerable injury by giving prizes for useless birds, and ignoring to a great extent breeds that would be of benefit to the farmer and the nation at large."

Mr. Edward Brown, a well-known English expert, whose specialty is table poultry, commenting upon the difference between the English and French show systems, says :—"English fanciers' shows are not established for the improvement of poultry in their economic qualities, but for fancy points. The judge regards as all important shape, size, colour, comb, legs, and general contour, and does not care whether the fowls are likely to make good table fowls or first-rate layers. In France the judging is exactly reversed. The points which denote economic qualities are looked for first of all, and then an examination made for externals. They know that birds which have special characteristics are best, either as layers or on the table, and thus they look out for these points, and breed to them."

Mr. Consul Gurney's Report on the Agriculture of the Cherbourg District, France, presented to both Houses of Parliament, ran as follows :—"The agriculturists of Western Normandy, having given up cereals, now get a very fair return for their capital and labour out of dairy-farming, poultry-rearing, and market-gardening, and London furnishes them with a profitable market for their butter, turkeys, geese, and poultry. Their fowls are carefully tended, and, having the free run of the grazing fields and the cider apple orchards surrounding the farmstead, add largely to the profits of the farm. Fostered by shows which are based upon the fallacious principle of breeding for feather only, a grievous mistake from a practical point of view, poultry-keeping in England has become too much of a fancy, benefiting only prize-winners and opportunist poultry-breeders catering for the newest fad in shape and colour."

These quotations could be supplemented to an almost indefinite extent from the fanciers and fanciers' press of both England and America, but I will content myself and conclude with the following from a pamphlet entitled "Poultry and Eggs for Market and Export," by a former expert to the Department of Agriculture, New Zealand :— "It must not be presumed that a bird that takes first prize at a poultry show is the best bird for the farmer. It may be the worst. Most of the pure-bred birds in this country are of the fancy class, and have been bred for showing only. They are deteriorated in useful qualities by in-breeding, by breeding from birds known to be poor layers, or weakly, because they show good points in feathers. The result is a loss in constitution; the birds themselves are subject to disease, and their chickens hard to rear. The egg-producing qualities have suffered even more. Many breeds once noted for laying have lost their good name through fancy breeding. Keeping back pullets from laying in order to increase their size is often practised by fanciers, and has a depressing effect on fecundity."

All the handicaps to profitable poultry-breeding as enumerated in the above extracts were well known, but the usual apathy of breeders allowed this state of things to continue, until the late William Cook conceived the idea of producing a breed with constituents calculated to stamp it as one of the very best for commercial purposes, and that while amenable to " improvement in appearance" by fanciers, would be by its colour and constitution, able to withstand all the effects for ill to which other varieties have been subject, and whether the big black fowl originated by him, and named after the town in which he then lived, was one of the best commercial, and despite exhibiting continues so, is one of the purports of this paper to show.

Of the many varied regrets indulged in by writers on the injury done to commercial poultry by hobbyists, the majority are to the effect that fancy and utility should, or could, be combined in such a degree that fowls, when put in the exhibition pen, could be awarded prizes on the same principle as that of other stock. A judge, when making awards in the Ayrshire cattle class, first looks for the points which indicate purity of breed, and as the commercial character of this breed is a big milk yield, the judge then looks for the development of certain organs which are associated with and indicate a liberal milk supply. The same in breeds noted for beef qualities; type is looked for as an indication of the purity of the particular breed, and then the great muscular development and big square frame on which to build this meat. Sheep are judged in the same way for wool or meat, certain indices pointing to a superiority in either, *i.e.*, commercial superiority. Not so with poultry, for, as has been frequently said, the champion hen, although perfect in all the points of her breed, may not be capable of laying an egg, while the first prize rooster may be useless in reproducing his species. This brings me to the "utility and fancy combined," and from much research in the vast tract of American poultry literature, there is not a doubt but the Yankees have to a great extent combined the two. Their judging standards are not of the arbitrary nature of those adopted in England, and which we faithfully follow here, with the result that although the American prize birds in many breeds could not win, judged by our standards, at the same time the purity of race is much more pronounced than with us, many of their breeds being pedigreed for over a quarter of a century, the Felch and other bloods being as faithfully perpetuated and as familiar in certain breeds of poultry as the "Bates" in English shorthorns, the above pedigreed strains being noted for their great laying qualities as well as their show-pen excellencies, while other breeds not only retain the economic qualities which accompanied them upon their introduction, but have improved in usefulness, while the same breeds under our system have been improved almost out of existence. Brahmas, which for their unproductiveness have now but few patrons here, are among the most prolific layers in America, following closely on the Wyandotte for popularity. Plymouth Rocks, which have scarcely a dozen breeders in this State, are only second to Wyandottes in that country in popularity, and form the bulk of the American export poultry trade to England.

There are a number of breeders in England who advocate that a system might be adopted of judging commercial qualities by appearance, as in other stock, the specimens whose appearance indicates the greatest laying capacity or the best flesh-formers to be awarded the prize; in other words, a combination of the beautiful from a fancier's point of view, and the economic from the commercial standard. Thanks, however, to the laying competitions incepted here, and now general throughout all the Australian States, it is being abundantly proved that under the present English Poultry Club standard as used in Australian shows, the utility and the fancy side of poultry can, and is, being combined, and that in more than one breed or variety which made laying records included specimens of the best exhibition type and colour of their respective breeds ; and there cannot be a doubt but that should these competitions continue, and patrons of the various breeds look after the appearance as well as the performance of their birds, ere long the combination of the fancy and utility side of poultry-keeping will be an accomplished fact; and so far as the breed which gives the title to this paper is concerned, for whatever or all purposes required, the performances and records already made conclusively prove that exhibition and useful qualities can safely be combined in one breed.

Chapter XVII.

Justification for New Breeds.

In the preceding chapters I have shown that the majority of the poultry productions of the past fifty years were heralded on their introduction as superior in some commercial essentials to their predecessors; that those who tried them, and the most competent authorities of the day subscribed them as superior, and just as sure in later years did the majority of these breeds, in an economic sense, come to grief, the show-pen system receiving credit for their retrogression. An article in a leading fanciers' paper on this subject I think worthy of reproduction :—

"Those who have much experience in poultry-keeping are aware that although the laying qualities of various breeds are capable of being described as good, bad, or indifferent, this description by no means universally applies to individuals or families of the breed. Even in breeds which are known as very good layers, some birds will be found which are far below par in this respect; while in breeds noted for their poor laying qualities, exceptional birds sometimes produce wonderful results.

"This divergence has its origin, we think, in the fact that birds are in this country bred from two perfectly distinct, and, to a certain extent antagonistic, objects. On the one hand, the poultry fancier is striving for perfection in colour and marking, for size, and for other fancy points. This aim he pursues quite without regard to the laying qualities of the birds selected to be bred from. It may be that the best fancy bird is a bad layer; but none the less the few eggs she lays are treasured and hatched out in preference to those of any other bird in the yard.

"This alone materially affects the laying qualities of many prize strains. In addition to this, however, in those breeds in which size is an object, exhibitors actually take means to prevent the early laying of the pullets. It is found that upon her commencing to lay, the growth of a pullet practically ceases for the time being. If this period can be delayed she continues to grow, and thus makes a better exhibition specimen. This is done by using non-stimulating foods, and moving the birds from yard to yard according as they show the least symptom of attaining maturity and commencing to lay. A still further reason for deterioration is to be found in the fact that the best birds are sent much about from show to show.

"This, like moving from place to place, but to a much greater extent, retards laying. These processes being repeated from generation to generation, the laying qualities of almost all prize stock have been materially impaired; and it is only in exceptional instances that birds of prize strains are good layers.

"It is curious to note how, time after time, the latest novelty in the poultry world is hailed as the layer, and how, time after time, the laying qualities are gradually lost, and the supposed first-class layer loses its character in this respect even amongst its warmest votaries. As an illustration of what we mean, we may instance the Brahma and the

Leghorn, both of which when introduced to this country had claims, and just claims, for pre-eminence as layers. Now the Brahma has almost universally lost its character as a layer, while the Leghorn is just hovering between the character of a good layer and a bad layer, according to the particular strain which is kept. On the other hand, there are throughout the country a few (though far too few) persons who regard laying qualities as of primary importance, and who, in the breeding of their stock, carefully keep that end in view. One breed or another may be selected for this purpose, and as purchasers from these strains find their birds are distinguished by good laying qualities, they rush to the conclusion that others of the breed are equally good in this respect. In this way the greatest diversity of results is obtained from different individuals of the same breed, and we see the curious spectacle of one poultry-keeper writing to say that he can only get eighty eggs a year from his Andalusians fed in such a way, while another hastens to reply that he keeps the same breed and gets 180 eggs per annum.

" Poultry shows have undoubtedly done good in establishing the pure breeds throughout the country ; but poultry-fancying does harm, in that it casts abroad through the country strains of inferior laying qualities, and thus checks in many instances poultry-keeping on a small scale without any view to exhibition. The difficulties pointed out in this article materially affect the poultry prospects of the county."

The majority of people, realising the truth of the above, might very naturally be expected to condemn a system involving such disastrous results to the varous breeds. However, the fact remains that as deterioration has given the cue to some enthusiast who shortly appears with a new breed possessing the strongest of all claims to public recognition, the great handicap to all the hitherto productions being the fact that a universally good breed could not be kept an eternally good breed. This great drawback to the best breeds was to many of the poultry-keepers well-known, and more than one attempt was made to produce a fowl combining an excess of meat and egg properties, and adapt itself to the fanciers' art, without involving any decadence in usefulness. This ideal fowl, or at least, as near so as can within reason be expected, was at last produced, and from the time it was sent out to the public in 1886 to the present day, has not only become the working-man's and the rich man's fowl, but has increased in popularity, and although taken in hand by the fancier, has not, like other breeds, deteriorated in useful qualities, but continues the same good all-round fowl as it left the hands of its originator, who named it after the village of which he was resident—Orpington (in Kent).

Originator of the Orpington.

Prior to 1883 poultry interests in England had no special journals, as now. Poultry show reports and poultry interests generally were looked after by the *Stockkeeper* and the then *Live Slock Journal* ; but as dogs, pigeons, and other fancy stock found a place therein, little space was allotted to the fowls. A then, and yet, very popular weekly, *The Journal of Horticulture*, conducted by Dr. Hogg, LL.D., gave weekly insets devoted to poultry, amongst its contributors being the late A. Comyns, LL.B. The success of this section of the journal prompted the proprietor in making a new venture in the fanciers' world by starting a paper entitled *Poultry, Pigeons, &c.*, of which Mr. Comyns became editor. About, or rather, before this time, a good deal of correspondence had appeared relative to a man named Cook, who was making a great noise in the poultry world—not as an exhibitor or prize-winner, but relative to the keeping, breeding, and crossing of fowls for profit; and not content with that prominence, he actually published a book entitled "Cook's Breeder and Feeder; or, How to make Poultry Pay," many of the statements contained therein being of such a glowing nature relative to the possibilities of the business that he was ridiculed by the exhibiting poultry breeders.

Mr. Comyns, the editor of the new paper, on its introduction stated that the practical side of the poultry question, as apart from the fancy, would be a subject of much importance in its columns, and to that end he had carefully gone through Mr. Cook's book; and although the figures therein were startling, and the results of certain crosses also wonderful, the book was favourably reviewed; but to further test the accuracy of certain statements, Mr. Comyns visited the author at his then residence, Chislehurst, the following being his report:—

"Hostile criticism of Mr. Cook and his book, and incredulity as to his statements, had not been wanting, and we endeavoured, so far as lay in our power, to so frame our inquiries and make our investigations to test the accuracy or inaccuracy of the statements contained in the book, and the amount of credit to be attached to the author's statements.

"We must state, in the first instance, that Mr. Cook is a working man in the strictest sense of the term. He holds a situation in the employment of a gentleman, and resides himself, with his family, in a small house or cottage—one of a row of similar houses. He has been for years past a breeder of one sort or another of live stock, and has devoted special attention to the production of fowls remarkable for their laying qualities."

Mr. Comyns showed that in Mr. Cook's small yards there were a number of breeding pens, the male birds in each pen being mated to hens of a different breed, and consisting of Dorkings, Houdans, Game, Brahmas, Hamburgs, Minorcas, Cochins, and Plymouth Rocks. It will be noticed that one or two of the most popular varieties—Wyandottes and Orpingtons—are absent from Mr. Cook's list of crosses; but it must be remembered that Wyandottes, although then in England, were in such limited numbers, that in the year 1882 not a single show provided classes for that breed; and although a few specimens did appear in show pens, they were only to be found in that refuge for the destitute "any other variety" class, hence their absence in a small yard in Chislehurst need cause no surprise, while for Orpingtons, they were only issued to the public in 1886. Still, as Mr. Cook acknowledged several years labour in their manufacture, the progenitors of our Orpingtons were among the list of breeds or crosses then located in Mr. Cook's back garden, on the occasion of Mr. Comyn's visit as reported above; and whatever ambition or hope Mr. Cook then entertained in the way of making a new breed of black fowls, the name under which they became known could not then have entered his head, seeing that they were called after the village to which he removed in after years.

After the interview alluded to Mr. Cook became a regular contributor to the columns of *Poultry*, his articles being original and practical as well, and were highly appreciated by those breeders whose object was profit. Cook's blunt statements, new methods, and astonishing figures frequently provoked sharp criticism, but these affected the new poultry apostle for nought. He went on his way writing, lecturing, and otherwise propagating the then new doctrine of making a living from poultry. Cook's name was now becoming a household word amongst economic poultry-breeders, his small place and methods being always free to those who came to look for information. Nearing the close of 1883, two well-known fanciers visited his yards, and reported as follows :—" We were not at all prejudiced in Mr. Cook's favour, but rather the reverse. The appearance of the crossbred birds is not pleasant to a fancier's eye, lacking that uniformity of colour which is so desirable. What a pity it is we cannot combine the good, the useful, and the beautiful. We inspected the result of thirty crosses, but those that struck us most were Houdan-Minorcas, a square-shaped, short-legged, with rich black plumage. We were also shown a fine collection of eggs from the crosses, rivaling in size the Andalusian, and there were any number of them, for box after box was produced.

About July, 1883, Mr. Cook removed from Chislehurst to Tower House, Orpington, where he had more room for his poultry experiments, the increased accommodation and better breeding facilities affording him much more extensive experience, which was readily and freely given in contributions to the journal already referred to. One of his articles—Pure Breeds *v.* Mongrels—I think worth repeating here :—" What I want the people to understand is that birds may be pure-bred without having their fancy points developed to such an extent as in the case of exhibition stock, and that these pure-bred birds may be bred with special regard to their useful points without allowing them to degenerate into nondescripts. I can assure my critics that it can be done, and that useful qualities so bred for can be established in pure breeds with far greater certainty than in any mongrels I have ever seen. Uniformity of appearance must also count for something. Those who bred for exhibition have done much good in the way of improving size, and those who go in for first crosses have done much in the way of meat and increased egg production. I do not say but that there are some decent birds to be found in the yards of mongrels; but I have never heard of anyone's yards being improved by these birds."

This, then, is a brief review and opinions of the man who a few years later put to the public a new breed of fowls of large size, combining the best qualities of the most desirable breeds, and one that inbreeding with its deteriorating tendencies need not be resorted to by fanciers to improve it for the show pen.

CHAPTER XVIII.

ORPINGTONS.

Origination.

ALTHOUGH W. Cook had been breeding poultry from 1870, and was in a limited circle fairly well known, it was not until 1882 that poultry breeders generally became aware of such a man. His various contributions to the then weekly poultry journal, *Poultry*, and the publication of a book entitled "Cook's Poultry Breeder and Feeder," brought his name into greater prominence, which, with other causes, may be expected to retain for all time wherever domestic poultry are known. During the years from 1870, Cook was experimenting by breeding, cross-breeding, and inter-breeding several of the then known varieties of poultry with a commercial object, but few were aware of the actual circumstances until 1886, when he sprang on the English poultry world what he called a new breed of fowls. These had black plumage and single combs, and much like the then exhibition Langshan, but with shorter legs and free from feathers. These he called Orpingtons, from the not very prominent Kentish town where he at that time resided. The announcement about them appeared in the two or three journals devoted to the fancy, the reception they received being sufficient to daunt any less determined man, and oblige him to abandon the thought of successfully launching in England a new breed of fowls, Yankeeland being considered to have the monopoly of poultry creation.

The papers and fanciers alike attacked both Cook and his fowls. The birds were alluded to by some as the ordinary black farm-yard fowl, by others as clean-legged Langshans, while those who wished to flavour their remarks with a little sarcasm, denominated the birds as Cook's commoners.

The unmerciful criticisms were all taken in good part by the Kent enthusiast, who believed he had manufactured a good commercial fowl, suitable for the back-yard of the city breeder, the road-side rearer, the extensive range of the farmer, and worthy the notice of the fancier, who, like the patrons of the turf, are ever profuse in the justification of their hobby on the grounds of improving the stock. However, I prefer to give Cook's own words relative to his experiments, investigations, and intentions, which culminated in what undoubtedly is one of the most popular and profitable breeds of domestic poultry :

55823   E

"Many years ago I conceived the idea of adapting poultry-keeping to present-day needs, and after revising the methods of feeding and general management, I found that the old breeds were, in many instances, sadly behind these times of scientific investigation and methodical systems of economics. Well do I remember each successive step, and each oft-recurring success, until the sneers of the old breeders, and the opposition of the few, faded into insignificance. A little opposition will be no great hindrance to the furtherance of my great purpose, and I intend always to further it until it be a complete and crowning success. One marked feature of the Orpington has always amazed me as much as it has gratified me, that is, the rapid development and improvement of the breeds. Some may be interested to know to what cause I attribute this rapid development. First, I may mention that in the making of the Orpington, good stock only was used. The birds then came forth excellent, not only in formation, but in material, out of which the breed was formed. The best blood brings vigour and stamina into the birds which nothing else could produce. Then, again, the crossing of the breeds has imparted strength, which the admixture of various breeds has always produced; and the best blood blended upon right principles, of course produces even more excellent results, as each successive cross has left its mark upon the breed. Consequently, we can but foresee not only a great future for Orpingtons, but a greater breed of Orpingtons for the future," adding " the show pen will always possess a strange fascination for many—the fancy will always hold its votaries in obedience; but the great future of the poultry-keeping industry does not lie here, but in the hard-handed, hard-headed toiler on the land—the farmer, the fruit grower, the artisan, and the dwellers in the cottages of the country side—and in these are to be found the material out of which the new order of things is very largely to be evolved."

When the late W. Cook first put Orpingtons on the market, he gave the full history of his experiments, leaving nothing to conjecture relative to the breeds from which he evolved the new fowls. He told the public that he looked for the most suitable material, that is, birds of good breeds and of good laying strains, table qualities, vitality, and constitution. He says:—

"I took a good, black Minorca, which variety are extraordinary layers, black plumage, not putting on fat readily, with white flesh and skin of fine texture, but with black legs and tremendous activity. The points I hoped to procure from this breed were black plumage, red face, and large comb and ear-lobes. I procured the finest-bodied cocks of the breed that I could find throughout the country that had red ear-lobes. These had been killed or thought little of before, because they had not white ear-lobes. With these Minorcas I mated some Black Plymouth Rock pullets, which are 'sports' from the American Plymouth Rocks, owing to the Black Java having been used in the making of the Plymouth Rock breed. Plymouth Rocks are hardy birds, winter layers of a brown or tinted egg, and were thought a good deal of when they were introduced, although their yellow skin and legs have always been against them as table fowls in this country. These varieties when crossed produced black pullets and good shaped birds. The cockerels came of a mixed straw colour, and were of course useless for my purpose. With these beautiful pullets I mated a good Langshan of the old short-legged type, and, as is well known, these birds were extraordinary winter and particularly late autumn layers of deep brown-shelled eggs. They lay when eggs are scarce, more so than at any other time of the year. They are also fine-bodied, black birds, showing an iridescent, metallic-green sheen upon their magnificent plumage. So these birds, with their long, deep breastbone, and white skin and flesh, infused many good qualities into the breed. The feathers down the legs had to be disposed of, and to do this I sought out all the Langshans that were produced without feathers on the legs. These Langshans laid earlier than the feather-legged ones, and, with their breasts curved in that peculiar way which denotes strength of constitution and their fine qualities, they soon set their stamp of real excellence upon the birds. So with the 'wastrels' from good breeds I formed the Black Orpingtons, using birds that represented the poultry of the three continents, viz., Minorcas from Europe, Langshans from Asia, and Plymouth Rocks from America, and three varieties that were looked upon as three of the best for general purposes that England possessed. And the Orpingtons have now been tried and found most valuable birds in the coldest and bleakest parts of America and England, Russia, Holland, and Germany, so that, with Africa, India, Australia, Malta, New Zealand, and Norway, Orpingtons have now spread to nearly every part of the civilised world. In the most trying hot climates, where other birds have died, Orpingtons are thriving and doing well."

This then is how and of what the new ideal fowls were made, and it is but a moiety of the tribute to the originator to acknowledge that of the many breeds and varieties of domestic fowls now extant, there is not one of them that in such a short space of time became so universally plentiful and popular, the breeders generally admitting that the new fowl fills the vacancy which existed in the poultry world—a good table fowl with white skin, a prolific layer of brown eggs, of a good appearance, easy to breed and rear, and fast growers. Cook's own description of them, written eighteen years ago, is as follows :—

"The hens are good sitters and mothers, but not so troublesome as some of the other varieties when not required for that purpose. They are free from feathers on the legs, and have a beautiful black, glossy plumage. They stand confinement, and lay through the severest of wintry weather, producing a beautiful brown egg of a fair size. If required for table they are very satisfactory, being an excellent flavour, and the breast meat remarkably white. The birds fatten up quickly for table when young. They breed true to type, though occasionally some of the cockerels may come with a little red in their hackles or feathers on their legs. The plumage is very glossy on both sexes, but more particularly on the cock. The sheen should be much the same as that of a good Langshan, single comb, evenly serrated in both sexes, standing erect in the cock, not large but very neat, very deep and long in the breast, red face, black legs, white toe-nails, four toes on each foot well spread out. The male birds should always be fully developed in the points where the hens are deficient. It is always best to mate up unrelated strains. There is nothing like fresh blood, if healthy, vigorous, and good laying birds are required. The Orpingtons have been known to lay from thirty to sixty eggs without missing a single day. The pullets are in full lay at six months old, many of them at five months. They are splendid birds to breed from, no matter whether the soil is wet or heavy, and they do well when they have a free range. At the present day no breed of fowl is used more for crossing with mongrel hens and pure birds of other varieties than Orpington cockerels."

I have already said and shown that the fanciers in England received the new breed with anything but enthusiasm. Mr. Cook was not a fancier in the general acceptation of the term, neither was he an exhibitor; and any poultry production emanating from any source but these was regarded with suspicion; hence when the single-comb, clean-legged black fowl was spoken or written of as a "new breed," the matter was pooh-poohed, one contributor to the papers at that time going so far as to describe them as an imposition, and now at the close of a nineteen years' experience, and at the height of popularity, these same fanciers are wondering at themselves for their then opposition, and this opposition can be rightly wondered at, seeing that but a few years before the English fanciers received with open arms from America, first Leghorns, then Rocks and Wyandottes, and although the two latter breeds were justly spoken of "as the best all-round fowls," still, for commercial purposes in England, both had the handicap of yellow legs, with the too frequent accompaniment of yellow skin, while the skin of the abused Orpington was as white as that of the much-favoured average Dorking.

A recent writer on this subject of English home manufacture thus puts the matter :—" It is from America that most of the new varieties of fowls come. American efforts in this direction have been by no means wanting in success. Witness the several sub-varieties of Leghorns, the now popular Plymouth Rock, and Wyandotte. An English fancier occasionally makes an effort to strike out in a new line, but whatever may have been the case some forty years ago, when the fancy was comparatively in its infancy, there is no doubt that now-a-days, the attempt made to bring forward new breeds of home manufacture do not meet with much encouragement. From the point of view of the fancier, there is plenty of room for new varieties, especially in the utilitarian sense, provided they be distinct and sufficiently established not to betray the original elements from which they were manufactured. It is probably on account of the premature way in which the results of English attempts to form new breeds have been brought forward that they have failed to attain anything like the success of imported specimens of manufacture. We have also a national prejudice in favour of being imposed upon, and just as we accept American bacon more readily when it is labelled 'Best Limerick' and sold at a high figure, than when it is truly described and priced accordingly, so we much prefer to be told that our new varieties of poultry are descended from birds imported from some unknown region, rather than learn the truth as to the elements from which they were formed."

In relation to the Orpington opposition, the truth of the closing sentence of the above quotation will be apparent, and there is not a doubt but had the originator maintained an air of mystery in connection with the component parts of his new fowl, its popularity would have been earlier assured. However, this opposition was little heeded by Cook ; he continued on his course of poultry-breeding for profit, lecturing, and otherwise propagating the new poultry gospel, advocating improved methods in the industry, and pleading with his audiences and the public to give the new breed a trial ; and as the majority of those who attended his lectures did so with the purpose of receiving instruction in the way of better-paying results, and to that end were influenced in favour of the new breed, and of the many hundreds of letters published on the result of these trials in the early history of the Orpington, the vast majority was in laudatory terms of its superiority. These unsolicitated testimonials from ordinary poultry-keepers to the merits of the new breed, soon brought conviction to the fanciers that there was something in them after all, and as the fancier who goes in for a new variety in its early history usually makes the most money out of it, so it was with the Orpington ; and as an occasional setting of eggs or trio of these birds from Cook got into a few fanciers' hands, there was soon application made to the societies for classes for the new production. Their appearance in competition for a year or two was confined to the variety class, but their rapid spread throughout the country amongst all classes of poultry-keepers, obliged the societies to make provision for this variety in their schedules, and they have now for a number of years become a recognised breed with agricultural and fanciers' societies alike. The Orpingtons already referred to are of course the Blacks, and the character given to them by the originator and testified to by thousands of breeders almost precluded the thought that any further inventions in the way of breeds or varieties would ever again come from that quarter. The birds were as near perfection as reason could expect, and were, as Mr. Cook said, able to withstand the hitherto ill-effects of the show-pen. However, the originator, seeing the popularity which his birds were attaining in the exhibition world and the fancy price which show-pen specimens realised, in addition to his utility propaganda, became an exhibiting fancier as well, and once he entered upon this course, it was no surprise to hear that other varieties of Orpingtons were in course of manufacture. These have duly appeared, and like the Blacks became widely distributed, one colour at least—the Buffs—for a time threatened to overshadow the originals ; however, the past two or three years, patrons of the Buffs are on the decline, and there is now scarcely a doubt that if Orpingtons continue for all time to occupy a place in domestic poultrydom, Blacks will be found in the chief position. The Department's artist has been happy in portraying a typical Black Orpington cock in the September

Gazette, and a hen of the same breed in the October number, these and the standard for judging, which appear later on, will enable the breeder to realise how near exhibition requirements their stock is, and perhaps save the novice from the disappointments of the show-pen, by enabling him to detect the shortcomings of his fowls in that respect.

CHAPTER XIX.

White Orpingtons.

THE success of Cook's first experiment in fowl-making, as can be readily imagined, prompted him to further efforts, which resulted three years after the production of the Blacks, he, this time, going to the other extreme in colour; the due arrival of White Orpingtons in 1889 fairly astonishing the poultry fraternity. The explanation, or apology for the Whites is not very definite, but is embodied in the following :—" I proceeded to produce, as far as possible, an ideal set of breeds that would do great things for the poultry-keepers of the present and the future," consequently Whites are one of this set, and the method of their production is as follows, in Mr. Cook's own words :—

"I had made many experiments to find out the best method of producing white fowls that were up to date. I began by crossing the White Leghorn cocks with Black Hamburg hens, and the pullets from this cross came every one white. I next used a rose-comb White Dorking cock, mated with the offspring from above, and some of the birds came blue, some barred, like Cuckoo Dorkings, and it was a long time before I could produce pure white birds. But with careful breeding, the rose-combed White Orpingtons were made into a breed which produces quick-growing, vigorous birds, with good laying and table qualities; they have taken wonderfully, and poultry-keepers have found them lay more than any other white breed—245 and 250 eggs each a year some specimens have produced, and others have laid 190 eggs in the same time. They are splendid table birds, have white skin and legs, and this is a point with English people. Then the single-comb White Orpington was produced by using a single-comb White Dorking in place of the rose-comb. There is the same difficulty with these birds as there is with all other white varieties—that is, although they do well in confined runs in town, the plumage shows the dirt, and never looks bright. As layers they can hold their own with almost any other breed, the eggs being of a cream colour, and of a good size. The chickens grow remarkably fast, and feather as quick as young partridges. The cockerels can be killed at an early age, and their skin is as white as that of a Dorking, and the legs the same colour; the pullets will lay at five months old, but most of them are in full lay at six and a-half months old. The cocks are splendid for crossing with White Leghorns, Light Brahmas, or White Dorkings—that is, if the owner wishes to keep his stock birds white, and yet does not at the same time wish the laying qualities to go down. It must be understood that new breeds are usually excellent

Orpington, and sure enough it led on to fortune. The Blacks were exactly nine years on the markets when Cook placed the first pair of Buff Orpingtons in the Dairy Show in 1894, and so great was the run on them, that before the end of the season he had disposed of 400 settings of eggs, and in the first year sold over 200 stock birds of this variety.

Writing a year later on the subject, Mr. Cook said :—" No variety of fowls up to the time Blacks were brought out ever took so well in England as they did, but the Buffs have sold off quicker in the time than even the Blacks. Some people may say : Why bring out the Buffs when the Blacks do so well—is it not overdoing it ? Not in the least. No one ever complains of a florist bringing out new flowers, particularly if the colour is more attractive than that of the old varieties ; so it is with the Buff Orpingtons. In introducing the Black and White varieties, I told the people how they were made, and I will do the same with the Buffs. First I mated a Gold-spangled Hamburg cock with a coloured Dorking hen of good size, and from the best laying strain obtainable; these produced pullets of a reddish-brown colour, which I mated with a Buff Cochin cock. Hamburgs are excellent layers, but rather delicate to rear as chickens. The Dorking is a splendid table bird with a very long breast, and when crossed with the Hamburg will produce wonderful layers."

As already said, Buffs were brought out in 1894, the originator antici-pating such a demand that in 1895, 1,000 chickens of this variety were reared at Orpington House, the craze for them being that long before the close of the year every bird for sale was disposed of, the majority of them being dispersed throughout England, large numbers of them also going to South Africa, Canada, United States, New Zealand, and Australia ; while in the early months of 1896, foreign orders to the extent of £1,000 were received. This extraordinary demand I have attributed to a craze, for the simple reason that the time this variety had been before the public was insufficient to warrant any definite statement as to their merits. However, as time went on, testimonials increased, and by the end of 1896 they were overtaking the Blacks for popularity. The extraordinary rush on or for this new variety being even a surprise to Mr. Cook himself, who in attempting to explain such, said :—" One reason why Buff Orpingtons have taken so well this past season, is because of their quick growth. This got talked about, and was an inducement to many people to go in for eggs who would otherwise not have done so. Buff Orpingtons, besides being good layers, are excellent table birds, and although they do not all come true to colour for exhibition purposes, they can be sent to the London market or utilised for one's own table. Another reason why so many took to them was because they were so good winter layers. In Sussex and Surrey they are used very much for turning down in farm-yards with mixed lots for improving the laying and table qualities, and these have done good service in this direction, and where birds are not perfect specimens, are very useful for the purpose mentioned, and are used largely by poultry-keepers for improving their stock, as they can be bought at a low price and are as good as the best for that purpose. Farmers and others have done very well by using them as they have, because the pullets have laid before they were six months old, and gone on for a long time, thus proving the laying and table qualities."

Earlier in these articles I have referred to the spleen shown by a few of the old fanciers against the Black Orpington in its early history, but as this had almost disappeared on the introduction of the Buffs, it may be thought that they, realising the triumph of the former variety, would pause before uplifting their voice against the latter; but this was not to be, all the old rancour was revived, while prominent breeders who had allowed the Blacks to come and conquer without resent appeared in force against the latest innovation—the Buffs. This policy was rightly described by Cook as the violent and virulent opposition of the old-time enthusiasts, who, dog in the manger like, would rather waste opportunities than let others, better capable of supplying birds for the times, carry on their benevolent work, and when the new breed appeared an outburst of nonsense hailed their introduction until he had to meet arguments which were as foolish as they were spitefully ingenious and as harmful as they were erroneous. How far they succeeded the popularity of all the varieties of Orpingtons to-day is a fitting reply, and, so far as the Buffs are concerned, Cook's prophecy in 1898 about them becoming as numerous as the Blacks was first overwhelmingly demonstrated at the Dairy Show in 1902, when black cockerels numbered 35 entries and pullets 43, buff cockerels had 47, and pullets the extraordinary number of 73.

In England, and of late years in a small way here, when any breed of fowls has become at all plentiful among fanciers a club is usually formed to assist in developing it by providing special prizes, securing competent judges, and otherwise framing regulations in the interests of the breed concerned. These clubs when conducted with discretion are of much value, but when personal interests or dictatorial conduct of any kind is allowed to creep in their usefulness is to be questioned. Black Orpingtons, like other varieties of fowls, had soon an organisation of this sort formed by a few of its earlier admirers. At an annual meeting which was held at the Holborn Viaduct Hotel, London, on the 20th January, 1895, there were eighty-four members on the roll. The Buffs had then been a year before the public, and the recruits to this new colour, thinking that the club could much enlarge its usefulness in embodying the Buffs in their constitution, made a proposal to that effect, but such was not to be, the following paragraph appearing in their annual report:—"The club view with strong disfavour the introduction during the past year of the so-called White and Buff Orpingtons, which are not allied to Orpingtons in any respect whatever." And before the meeting closed the following resolution, moved by the president, General Gillespie, was put and carried:—"That the club view with dissatisfaction the providing of classes by show committees for other than Black Orpingtons, and the club hereby refuse to grant medals or specials for any than the variety recognised by it." And, as showing the tyrannical nature of some of these clubs, a 10-guinea special challenge cup was offered by the president and accepted, one condition of the competition being "that it was only open for competition to those who had never exhibited Orpingtons other than black."

The resolution adopted by the meeting in question much affected Mr. Cook, who took umbrage at the club referring to the new variety as so-called Orpingtons, and in the following issue of his poultry journal gave his opinion on the attempted boycott:—"As I am the originator of the Orpington fowl," he begins, "I think it only fair that I should be allowed to make a few remarks upon the report. When I bring out a breed of fowls, I always let the public know what blood is used to produce it. When some Americans bring out a new variety they give the best side of the birds, saying nothing about whether they breed true to colour or have any other failings—they leave the public to find that out for themselves. What first induced me to bring out Orpingtons was seeing the Americans bringing out new breeds or varieties, and the English people taking them up, in many cases paying very large sums of money for good specimens. I could see no reason why some of that money should not be kept in England —in other words go into the Englishman's pocket as well as the American's. I knew quite well we had many breeds in this country which could be blended together to produce a new variety as well as the Americans, in fact, we have better breeds to select from, for the Americans have never brought out a fowl equal to the Orpington, in appearance, winter laying, and table qualities combined. As I had been trying experiments for many years in crossing and re-crossing, I knew exactly what breeds to use to produce such birds, and I am pleased to say that my efforts have not been in vain. I do not say a word against any of the breeds the Americans have produced, but I do want to ask why we should pay such enormous prices that in many instances have been paid to the Americans, when we can produce quite as good, if not better, ourselves. No one complains that we have too many varieties of flowers, yet we are constantly having fresh sorts introduced. As long as they are pleasing to the eye, and superior to the old ones, people admire them, and buy them at the same time. It is only a few old breeders who have not yet got quite out of the roads our grandfathers travelled in, and some others who are jealous of the success the Orpington has met with, that speak against these birds. I do not wish to boast of my success, but I am thankful people have taken the breed up in the way they have done."

With the above plain statement of facts in 1895, and the since unprecedented success of the variety as already noted, it would naturally be supposed all opposition would have ceased. It, however, did not, for within another year or two a most acrimonious correspondence arose over the breed. The opposition this time going so far as to question the veracity of Mr. Cook as being responsible for the manufacture of the variety, they asserting that the breed was nothing else than an old buff fowl common in two or three English counties, and known as "Lincolnshire Buffs." Messrs. Lewis Wright, E. Brown, F.L.Z., Harrison Weir, and a number of other poultry authorities all joining in describing the birds as nothing else but an ordinary buff farm-yard fowl largely bred in Lincolnshire.

The correspondence on this subject would fill an entire *Gazette*, but having gone carefully through every line of the discussion, it is not difficult to arrive at a conclusion as to which side the preponderance of evidence rested, suffice here to say the continued popularity of the Buff Orpingtons, and their acknowledged excellent qualities is a complete justification for their introduction from whatever source; while for Lincolnshire Buffs, which were rarely heard of in the poultry press prior to 1894, they have now emerged from that obscurity which in the absence of the discussion they were destined to remain, and as Buff Orpingtons are now popular here, I have considered it right to briefly record the story of their rise and development, together with the formidable opposition they encountered in the early years of their history.

CHAPTER XXI.

Jubilee and Spangled Orpingtons.

HITHERTO the productions in the poultry line, which emanated from Orpington House, although of several colours, were but self colours, namely, blacks, whites, and buffs, and of these, two varieties each, single and rose-comb. These, however, did not satisfy the originator, who, ever fertile in resources in the way of producing type and colour in order to secure an ideal, now proceeded to invent another variety composed of several colours. Mr. Cook said, he had been making at these for a number of years, and first placed them before the public in 1897—Jubilee year—and named them Diamond Jubilee Orpingtons. Her Majesty, the late Queen, the same year adding a breeding-pen of this variety to her large collection of prize poultry. Mr. Cook wrote of them as follows :—" The rapid spread of the new breeds has been the wonder of the poultry world, and, as one after another of them has been launched and absorbed by the great army of breeders, and splendid classes, well filled, have appeared at the various shows, the voice of controversy, which at one time meant only the voice of derision, has changed into one of praise and admiration. Nothing succeeds like success, and Orpingtons have succeeded, but the Diamond Jubilee Orpingtons have succeeded in a way peculiarly their own. Those who have tried them as table birds, say they have never eaten such splendid fowls, and we are not surprised, as they have been bred in such a way as will ensure this, and they will become better if they are carefully bred. The laying qualities of these birds are also very extraordinary; some that were hatched in February were laying in August, while the March and April chickens were laying in early September, and these laid all through the winter, some averaging 100 eggs up to the end of February, by the time they were twelve months old. The chickens grow as fast as any other of the Orpington varieties, indeed, these and the Buffs grow, if anything, faster than the Blacks. We have had a run on the cockerels, as these are found especially valuable for crossing in the farm-yards, and this is a valuable feature of the influence these birds are introducing into the already improved condition of things among the farmers' fowls."

In a communication I received from Mr. Cook, in April, 1899, he stated that the Jubilee fowls were then very popular in England, on account of their wonderful laying qualities, and handsome plumage. The first Jubilees to reach Sydney was in December, 1899, a trio each, arriving to the order of Mr. Hayes, then of Epping, and a trio to the writer, one hen of the latter being still alive, now seven years of age. Following these were a shipment to Mr. Gillham. The above constituted the bulk of the stock in Australia until the following year, when the originator visited Sydney, bringing with him about thirty of the breed. This variety for two or three years did remarkably well, large classes of them appearing at the Royal and other Sydney shows; but during the past year there has been a serious decline in their patrons. Strange to say, however, in Victoria they have quite a number of admirers, and are spoken of there as excellent for the export trade which, as in other parts of Australia, still hangs fire. When the Jubilees were produced many thought the Orpington family was complete. Not so, for on Mr. Cook's visit to Sydney together with the Jubilees, a number of a spangled variety arrived. These were black and white, much like the Houdan in colour. The birds found new owners, but were a less success than the Jubilees. This then, closes the historical portion of the Orpington fowls, but not all the Orpington productions, waterfowl, like the other poultry, being honoured by the addition of a couple of new braeds to their rather limited family. The Blue Orpington Duck was put on the market in 1895, and reached Sydney a year later, and the Buff in 1897. The only reference of importance in connection with these is confirmatory of Mr. Cook's statements as to their merits. The Editor of *Poultry*, early in 1898, visited Orpington House, and in his report said :—" Of the ducks mention must be made of the Blue and Buff Orpingtons, two varieties made by Mr. Cook in recent years : the former combining the necessary qualities for table purposes, of quick and large growth, together with being good layers, while the Buffs are specially valuable as layers. These ducks have 'caught on,' as the saying is, and numbers have been exported to the colonies."

The photograph of the Orpington duck, taken by Mr. Grosse, early in 1899, and which illustrated the Orpington article at that time, shows the great size spoken of by the originator, while the records made at the two Duck Egg-Laying competitions testifies Mr. Cook's statement that the Buffs are specially valuable as layers.

CHAPTER XXII.

Judging Orpingtons.

HAVING now shown and traced the history of the Orpington from the time the late W. Cook conceived that there was a vacancy for such a breed until the last variety originated, it is now opportune to give the type or shape, size, colour, and other characteristics which go to make perfect specimens of the breed. It may be said that the shape of a fowl or the colour of its plumage need not trouble the farmer whose consideration is eggs and meat for his household or the market. There is certainly some reason in the contention, but largely superficial, for just as was remarked when treating on Wyandottes, the most perfect specimens of the breed will, if not carefully mated, very shortly deteriorate, and become a flock bearing little similitude to their progenitors, for it must be recollected that all the outward qualities of the Orpingtons, and inward as well, were first the result of crosses, and the continuous selection year after year of specimens that possessed the desired qualities in the greatest excess. Then when all these were at last found combined in a certain family, standards based largely on the characteristics of this family, or even on a higher ideal were formed, and by these standards the birds when exhibited are expected to be judged, and whether the farmer exhibits his fowls at the local show or not, it will be incumbent on him that while his object is eggs and carcases, he must keep in view the fact that except thought and experience be brought to bear in the mating and breeding, his stock of Orpingtons will soon be that in name only, and in order to show the utility man, and the exhibitor as well, the special points and requirements of the several varieties of this breed, the standards issued by the British Poultry Club are appended. The excellent illustrations executed by the Department's artist will largely assist in the interpretation.

General characteristics of cock :—

Head and neck.—Head : Small, neat, fairly full over the eye, carried erect.
Beak.—Strong, and nicely curved.
Eye.—Full, bright, and intelligent.
Comb.—Single or rose. The single comb to be of medium size, erect, evenly serrated, free from side sprigs. The rose comb should be set straight and firmly on the head, full of fine work or spikes, free from hollow in centre, and narrowing behind to a distinct peak lying well down to head (not sticking up).
Ear-lobes.—Medium size and rather long.
Wattles.—Medium length and well rounded.
Neck.—Nicely curved, with full hackle.
Body.—Breast : Broad, deep and full, carried well forward, long straight breastbone.
Back.—Short, with broad shoulders.
Saddle.—Rising slightly, with full hackle.
Wings.—Well formed, and carried close to the body.
Skin.—Thin and fine in texture.
Flesh.—Firm.
Tail.—Medium size, flowing, and inclined backwards.
Legs and feet.—Thighs : Short.
Shanks.—Short and strong.
Toes.—Four in number, well spread.
General shape and carriage.—Cobby, and compact ; erect and graceful.
Size and weight.—Large. Between 9 lb. and 10 lb. when fully matured.
Plumage.—Close.

General characteristics of hen :—

Head and neck.—As in the cock.
Body.—Breast, back, and wings : As in the cock.
Cushion.—Small, but sufficient to give the back a short and graceful curved appearance.
Skin and flesh.—As in the cock.
Tail.—Medium size, inclined backward and upward.
General shape and carriage.—As in the cock.
Size and weight.—Large. About 7 lb. or 8 lb. when fully matured.

Colour in Black Orpingtons :—

In both sexes.—Beak : Black.
Eye.—Black, with dark-brown iris.
Comb, face, ear-lobes and wattles.—Red.
Shanks.—Black.
Skin and flesh.—White.
Plumage.—Black throughout, with a green sheen or lustre upon it, free from coloured feathers.

Colour in Buff Orpingtons :—

In both sexes.—Beak : White or horn colour.
Eye.—Red or brown, the former preferred.
Comb, face, ear-lobes and wattles.—Red.
Shanks.—White.
Plumage.—Any shade of buff from lemon buff to rich buff, on the one side avoiding washiness, and on the other side a reddish tinge. The colour to be perfectly uniform throughout, allowing for the greater lustre on the hackle and saddle-feathers, and of the wing-bow in the case of the cock only.

Value of Points in Black Orpingtons, cock or hen.

Defects.	Deduct up to. Points.
Defects in plumage and condition	10
,, head, 5 ; comb, 7 ; face, 5 ; beak, 3 ; eye, 5	25
,, breast	10
,, saddle or cushion and back	5
,, tail	5
,, legs and feet	5
,, skin and flesh	5
Want of shape	15
Defect in carriage	10
Want of size	10
A perfect bird to count	100

Value of Points in Buff Orpingtons, cock or hen.

Defects.	Deduct up to
Defects in head and comb	10
,, colour	35
Want of shape	20
,, size	10
Defects in legs and feet	15
Want of condition	10
A perfect bird to count	100

Serious defects for which birds should be passed.—Other than four toes ; wry tail or any deformity ; the slightest feather or fluff on legs or feet ; long legs, yellow skin ; twist or side spikes in comb, or comb over to one side ; yellow in legs or feet. *In Blacks.*—Any coloured feathers. *In Buffs.*—Any white, or much black in tail or flights ; legs any colour but white.

When the above poultry clubs standards were compiled and issued in 1901, the White Orpingtons had made no headway in England, and were not included in the standards ; however, during 1904, and the early part of this year, the Whites and Jubilees were receiving more attention, and, as is usual in England, when a breed is likely to catch on, a club is formed, a standard formulated, and encouragement given to exhibitors in the way of special prizes.

In the early months of the present year one of these was formed in England, called the Variety Orpington Club, which, shortly after its formation, drew up and published the following standards. Such is not yet ratified by the English Poultry Club, but there is scarcely a question but it will be when presented.

Colour of White Orpingtons :—

Both sexes.—Beak : White.
Eyes.—Red.
Comb, face, ear-lobes, and wattles.—Red.
Shanks.—White.
Skin and flesh.—White.
Plumage.—Pure snow white, with a good lustre ; free from any foreign colour.

Value of Points.

Head	10
Colour	30
Condition	15
Legs and feet	15
Size and type	30
	100

Colour of Jubilee Orpingtons :—

Both sexes.—Beak : White or horn colour.
Eyes.—Red or brown, the former preferred.
Comb, face, ear-lobes and wattles.—Red.
Shank and feet.—White or pinky-white ; a little horn colour not to be considered as disqualification for the present.
Toe-nails.—White or horn.
Skin and flesh.—White.
Cock.—Plumage : N.B.—The term " mahogany " in this standard to be taken as meaning " bright mahogany, not dark nor maroon in shade."
Neck hackle.—Mahogany, with black stripe and white tip ; the shaft mahogany, of same shade as feather.
Saddle hackle.—To match neck hackle.
Back.—To follow neck and saddle.
Breast.—Mahogany, with black spangle, and white tip ; the three colours well broken and showing in equal proportions, avoiding a ticked effect on the one hand, and a blotchy effect on the other.
Wing bow.—To follow hackle.
Wing bar.—Black.
Secondaries.—Mahogany, black and white.
Flights.— ,, ,, but more white.
Sickles and true tail feathers.—White, or black and white, or mahogany black and white.
Coverts.—Black, edged with mahogany and white tips.
Thighs and fluff.—To follow breast.
Hen.—Head and neck : To match cock, allowing for difference of sex.
Body, breast, and back.—Mahogany, with black spangles and white tips ; the shaft mahogany, of same shade as feather. The three colours well broken and showing in equal proportions, avoiding a ticked effect on one hand and a blotchy effect on the other ; the effect to be uniform throughout the bird.
Wings.—As body, with flights as in cock.
Tail.—To follow the cock.
Thighs and fluff.—To follow the breast.

Value of Points.

Head	10
Colour	35
Condition	15
Legs and feet	10
Size and shape	30
	100

Colour of Spangled Orpingtons :—

Both sexes.—Beak : Black, or black and white.
*Eyes.—*Brown.
*Comb, face, ear-lobes, and wattles.—*Red.
*Shanks and feet.—*Black and white, mottled as evenly as possible ; toe-nails, white.
*Skin and flesh.—*White.
The cock.—Neck hackles : Black, with white tips.
*Saddle hackles.—*Black, with white tips.
*Back.—*Black, slightly ticked with white.
*Breast.—*Black, with white tips ; the two colours showing in equal proportions, avoiding a ticked effect on the one hand and a blotchy effect on the other.
*Wing-bow.—*Same as back.
*Wing-bar.—*Black.
*Secondaries.—*Black and white.
Flights.— ,, but more white.
*Sickles.—*Black with white tips.
*Coverts.—*Black with white tips.
*True tail feathers.—*Black and white.
*Thighs and fluff.—*Black, with white tips.
Hen.—Head and neck : Black with white tips.
*Body, breast, and back.—*Same as the breast of the cock ; the effect to be uniform throughout the bird.
*Wings.—*As body, with flights as in cock.
*Tail.—*As in cock.
*Thighs and fluff.—*As in cock.

Value of Points.

Head	...	...	...	...	...	...	...	10
Colour	...	...	...	...	...	...	...	35
Condition	...	...	...	...	...	...	...	15
Legs and feet	...	..	..	...	...	...	...	10
Size and type	...	...	...	...	...	...	...	30
								100

Defects :—

Feathers on legs ; long legs ; poor shape ; much white in lobes.

Chapter XXIII.

Orpingtons in Australia.

THE first Orpingtons to reach Australia were brought out by a friend of the late W. Graham, of Five Dock, and arrived on 27th November, 1887, and were the first Orpingtons to appear at any Australian show, namely Balmain, on the 4th July, 1889, Mr. Graham being the only exhibitor, their début at a Sydney show being made a fortnight later, there being a solitary entry. In no other breed or variety of fowls that ever came to Australia was there such a continuous increase in the entries annually in metropolitan exhibitions, the single pair of 1889 being now regularly represented by between two and three hundred. These exhibits, it must be remembered, are from breeders who have, or think they have, the very best specimens of the breed ; and when such a number of a breed appear at every show, it is evidence sufficient that Orpingtons are largely bred throughout the country, and indeed, was confirmation required of this, a visit to the poultry sale-yards any week would reveal the fact that the ordinary mongrel fowl, so long favoured in this as in most other countries, is giving place to pure breeds, Black and Buff Orpingtons holding chief places amongst the pure breeds which appear, followed by the Wyandottes. Nor does this exhaust the esteem in which the breed is held, for large numbers of the ordinary cross-bred market chickens claim an Orpington as one of their parents. Why this breed has become so universally popular can be best testified by my many correspondents on the subject. One prominent market poultry-man writes, "Plenty of eggs and a quick-growing carcase is what I required, and I gave up my favourite breed, the Plymouth Rock, for the simple reason that I found the Orpington the most profitable." Another breeder asserts that both Blacks and Buffs are the most profitable sorts he ever bred, and has handled Rocks, Wyandottes, and Langshans. A lady exhibitor says she has found the Blacks the most saleable of any fowls she has kept, and believes the Blacks are better than any of the other varieties. A country breeder mentions that the Black Orpingtons commence to lay before any of his other breeds, and also affirms as to their quick development. A number of other complimentary letters are before me, all exhaustive on the excellencies of the Orpington, and all conclusive that the breed is of all others the most popular, and is at the present time kept and bred on more agricultural farms than all the other breeds combined, Wyandottes excepted.

It has been already mentioned that the first to reach Australia came to this State, and it can be safely said that ever since the first arrival to Mr. Graham, and the second, by the ship "Macquarie" to Mr. J. E. Pemell, Sydney continues the headquarters of the breed, very large sums of money being received here from fanciers in other States each year for superior specimens for show purposes. Up to within a few years ago there was no official knowledge of the large numbers of this breed which each show season changed hands to Victorian breeders, but of late years, on account of poultry tick, a regulation by the Victorian Government demanding a clean certificate for each and every fowl that enters Victoria from New South Wales affords reliable statistics of the great quantities of Orpingtons from this State which each year appear at the Victorian Shows.

Further confirmation of Sydney being the great breeding centre for exhibition Orpingtons is confirmed by a report in *The Feathered and Kennel World* of a poultry show held in the Exhibition Building, Melbourne, on 23rd June, wherein it is stated that in a class of twenty-six cockerels the first, second, third, and fourth prize birds were bred in New South Wales.

In all the other States and New Zealand Blacks and Buffs are most plentiful, sharing in popularity with the Wyandottes, but for numbers and quality this State is ahead, the classes at the Metropolitan shows for cockerels and pullets often reaching nearly fifty entries each ; and that the Blacks will continue the most plentiful is generally acknowledged, the cause of their popularity being that they are considered for fancier and farmer alike the most profitable.

CHAPTER XXIV.

Orpingtons in America.

It is not surprising that Orpingtons became popular in Australia, seeing that they were made in England, where the overwhelming bulk of our prize poultry stock came from, and as the breed from its start had increasing admirers in England, we followed in the wake. It was, however, not a blind following, nor yet patriotism, seeing that the fowls actually possessed every attribute considered essential to a proper breed of fowls. These merits not only won them patronage here, but was actually a password of entry into every country in the world where domestic poultry are kept. For every specimen of the breed which came to Australia of late years, South Africa received hundreds ; they are now plentiful in France, Italy, Germany, Denmark, and other Continental countries ; while of the many millions of Russian eggs now sold annually in London, a fair proportion of them are laid by Orpington hens in Siberia. Coming to that country of big things—poultry farms included—it was thought by many that the breeders in the States would not take to the Orpingtons, first because they had not the yellow skin and legs which they so much desire, and second that it was not one of their own manufacture, as were the Rocks and Wyandottes. In my previous work on this breed six years ago extended reference was given to its prospects in the United States, the following extract sufficing for my present purpose :—" But whatever the prejudice in England in favour of white legs and skin, there cannot be a doubt about the American preference for yellow. Every one of the numerous and able poultry authorities of that country being strong advocates for bright yellow skin, whether for the roaster, the boiler, or the more youthful broiler; Brahmas, Cochins, Rocks, Wyandottes, and crosses from such being the varieties most largely bred for commercial purposes in that country. Consequently with this pronounced belief in bright yellow, it is not surprising that the Orpingtons found little favour in that great poultry-breeding country. However, when there is a good thing on the Yankees are soon found to be in it, and realising the increasing popularity of the Orpingtons in England during the past two or three years, several of the American fanciers have made very large purchases in England, the result being that many of the big American shows now make classes for this breed." Other references and comparisons between the English Orpington and the American Plymouth Rock were made, but all to the disadvantage of the latter. In due course the work reached America, where it received criticism from the poultry press of that country. One leading journal, *Farm Poultry*, devoted two pages to the work, and while complimentary to a degree on the merits of the book, was most outspoken when a suggestion was made that Orpingtons would become plentiful in that country, and severe in the extreme when the writer had the effrontery to compare an English-made fowl with their much cherished Plymouth Rock. The following from the critique shows the American feeling in the matter, and later references confirmatory of the prediction, that despite local or other prejudices, a good breed of fowl of whatever name or colour will win favour with practical breeders, irrespective of clime or country. The editor says :—

"'The Popular Orpington' is the title of a seventy-five page pamphlet by Geo. Bradshaw, Government Poultry Expert, just published by the Department of Agriculture, Sydney, New South Wales. It is a most exhaustive and readable pamphlet, devoted almost entirely to presenting the merits of this one breed, in its several varieties, to the farmers of Australia. Judged in the light of its purpose, the book is unquestionably one of the best publications of the past year. The subject is logically and skilfully treated, and but little matter is introduced that is not brought to bear well upon the points the author desires to make. I can easily imagine the Australian farmer overtaken by a worse fate than to have the conviction forced on him that it will pay him to breed Orpingtons. I can imagine, too, that were there no Plymouth Rocks or Wyandottes it might be worth while for someone in this country to devote himself to the task of persuading our farmers to use Orpingtons. As it is, the Orpington, if it is really a better fowl for the Australian farmer than the Plymouth Rock or Wyandotte is for the American farmer, is the best fowl there, for the same reason that it is not the best here, viz.—Because there yellow legs and skin are not specially prized, and here they increase the market value and economic importance of a breed. But when Mr. Bradshaw or any other admirer of the Orpington undertakes to show that it outclasses all the other breeds, he takes an untenable position. When he makes comparisons, quotes authorities, and gives facts and figures in support of his position, he challenges criticism; and when he expresses a very mean opinion of the Plymouth Rock as compared with the Orpington, or even the Wyandotte, which, in his judgment, is only a remote second to the Orpington, it seems to me he places himself in a position like that in which a certain well known judge of poultry once found himself.

* * * * * *

"As long as the Orpington enthusiasts content themselves with asserting that Orpingtons are good practical birds, having the same general characteristics as Plymouth Rocks and Wyandottes, and on some accounts more desirable than either for non-American markets, they are on safe ground. They are equally safe in saying that some Orpingtons are better than some Plymouth Rocks, or some Wyandottes. But when they begin to talk about Orpingtons as a breed being better than the other breeds, they must back their talk by more substantial, direct proof than anyone has, to this date, seen fit to offer. And when they predict for Orpingtons greater and more permanent popularity than other breeds have attained, they simply make the wish father to the prediction.

* * * * * *

" If the outline history presented in ' The Popular Orpington,' of the rise and fall of various breeds in Australia, teaches anything, it teaches that fashions in breeds of poultry are more fickle there than elsewhere.

The Australian breeder should learn from it that when one breed is at the height of popularity it is good policy to prepare, and prepare quickly, to fill the demand for its successor. If anyone there is at a loss to know what variety is coming next, he should communicate with some of our White Rock breeders. Considering the history of the varieties in Australia, and the recent history of the White Rock in America, that variety would seem the most likely one to 'take' next in Australia. In a previous pamphlet, 'Profitable Poultry Breeding for the Local and English Markets,' first issued in 1897, and now going through a second edition, Mr. Bradshaw's sole reference to the White Rocks is in this sentence: 'The ordinary blue Barred Plymouth Rock is too well known to be here described, and of late years a white has been produced, with little hope of a successful future.'

* * * * * * *

" Clearly the author of ' The Popular Orpington' has no use for the Rocks. In the paragraphs devoted to them in the pamphlet from which the sentence just quoted is taken, he is unwilling to allow that they are really good all-round fowls; speaks of their 'reputed' laying properties, and attributes their decline in public favour to lack of merit. If the Rocks which reached Australia lacked merit to maintain popularity in competition with the Wyandottes, and later with the Orpingtons, they must have been very poor Rocks.

* * * * * * *

"I would like to take Mr. Bradshaw out into the central west and show him the Barred Rocks on the farms where in State after State this is the popular variety. Nor is its popularity there of mushroom growth, or likely to prove transient. Throughout a quarter of a century it has gradually won its way into the confidence of the farmers as 'the business fowl of the nineteenth century.' It is to-day making headway more rapidly than ever, and the reputation of the Barred Rock has had much to do with the present phenomenal growth in popularity of the white variety, and the less general, but still marked rate of growth of the newer Buffs.

* * * * * *

"Mr. Bradshaw devotes a chapter to ' Orpingtons in America.' Considering the American preference for yellow-legged, yellow-skinned poultry, he thinks it ' not surprising that the Orpingtons found little favour in that great poultry breeding country.' However, he adds, 'when there is a good thing on, the Yankees are soon found to be in it, and realising the increasing popularity of the Orpingtons in England during the past two or three years, several of the American fanciers have made very large purchases in England, the result being that many of the big American shows now make classes for this variety.' Then he quotes letters from two American breeders, intimating that the popularity of the Orpington in America is assured, and closes by calling attention to records of shipments of Orpingtons to this country which appear from time to time in Mr. Cook's personal poultry journal.

"The Orpingtons should have an increasing number of admirers in this country, but that any variety of Orpingtons should ever become as popular as the Plymouth Rock is a stretch of imagination, and but father to the thought."

A great deal more was said by the American editor on behalf of the American breeds, and while not depreciatory of the Orpingtons, the whole tone of the article was to the effect that the English-made fowls would not make much headway in America. Whether such predictions or those of the writer's were the correct ones will be seen from the following, the facts being that the Orpington fowl from that time has been increasing in the States and to an extent far beyond any anticipations, so much so that two or three years ago a special monthly poultry paper was issued in its interests and entitled *The Orpington Fowl.* England, the birthplace of the breed, not being equal to that occasion yet. Thousands of pounds worth of stock were imported annually from England, the demand for the breed being such, that the late W. Cook established a large farm at Scotch Plains, New Jersey, for the special purpose of breeding these fowls for American purchasers, the latest report being that there are over 3,000 of young stock of this breed for sale. Clubs have been established for the encouragement of the several varieties, the last one being devoted to Buffs, and, as showing the progress made, within five months after its inception over 100 members were enrolled. The secretary lately wrote : " We cannot recall any poultry club which has made such progress in so short a time, there being hardly a State or Territory in the Union which is not represented in the membership."

It now remains to give the American opinions of the English breed, as written by some of its patrons in that country. Mr. Irving Crocker, a well-known breeder, contributes the following :—

" Although I have bred the Orpington fowl for two or three years only, I have been able to visit those who have bred them for a much longer time, and by asking questions kept myself pretty well posted since their first introduction into this country. From my own acquaintance with the breed, I can say that its good qualities have never been exaggerated. This was something of a surprise to me, for I always made some allowance for what I supposed to be the lively imagination of too enthusiastic admirers. But I have to confess, in the light of my own experience, that in doing so I did both the breeders and the breed an injustice.

" Right here let me say that in thoroughbred poultry I could never be satisfied with symmetrical proportions or perfect markings alone, much as I can admire them. I must have a breed that can be depended upon to fill the egg-basket, while the culls make fair returns as market poultry. This may be commercialism, but it is also business. I have never owned a breed that combined all of these traits to the same extent as does the Orpington, and I have experimented not a little. For beauty of outline they are certainly unsurpassed, while their rapid growth and early maturity is remarkable, to say the least. Added to these qualities, they are possessed of extreme hardiness. I do not recall ever losing but one from sickness, and that one was quite a young chick. Out of nearly 100 White Orpingtons hatched last spring, not one has ever shown the slightest symptom of disease. Some may say this was owing to the treatment they received. While that may be true to some extent, other breeds in the past, under the same conditions, were not exempt from colds, catarrh, roup, canker, bowel trouble, and other ills to which poultry flesh is heir. This being the case, I am forced to the conclusion that the Orpington has an especially strong and rugged constitution. I believe there are just two objections that are urged against this breed. The one most commonly heard is, they have white shanks. Now, this would he an objection worth considering if yellow shanks were an indication of any peculiar merit belonging to the breeds having them. So far from this being the case, we have but to compare some of the yellow-legged breeds to see that particular colour carries nothing with it but mere fancy. For example, we have the Indian Game, which is a good table fowl, but a poor layer. Then we have the Leghorn, which is a good layer, but worthless for the table. Added to these, we have others, breeds and no breeds, big and coarse, but of no particular account anyway, yet all with yellow legs. So this objection simmers down to a matter of preference, a question of taste, and educational prejudice. But there is something to be said for the white-skinned breeds as such. In the first place, they are exceptionally good layers. In the second place, this white skin carries with it a whiter, more tender, juicy, and delicate meat than can be found on any yellow-skinned fowl. So marked are these qualities that, in sections where the Orpingtons are grown, largely to the exclusion of other breeds, retail dealers in dressed poultry have a large and persistent demand for these fowls because of the qualities mentioned.

"I believe the time will come when the American breeders, becoming accustomed to these white shanks and more thoroughly understanding the superior merit accompanying them, will discard the prejudice that in some cases influences them, and admit that the white shanks are not so bad after all.

"Objection number two is 'it is an English breed.' I hesitate to mention this criticism because of its insignificance. The mere allusion is to give it undue importance. My only apology is that it has already been harped upon in the poultry papers. The class of would-be fanciers who offer this objection seem to be thoroughly tinctured with a bogus Americanism. They can see no good in any breed of foreign origin, and the favourable mention of an English breed drives them into a perfect frenzy of supposed patriotism, while they shriek 'Anglo-maniacs.' My advice to all such self-sufficient persons is to withhold their criticism of English breeds because they are English until they have produced one equally good."

Mr. Wallace P. Willett, of New Jersey, one of the first American Orpington breeders, says: "The surprise to me is that this wonderful breed, which has made its own way on its merits over all the rest of the world, has been kept in the background so long in these United States. There is not a country in Europe, Asia, Africa, or Australia where this breed has not taken the lead of all others, and it has only been brought out five or six years. Now that the United States is fully ripe and ready and on the look-out for a new sensation in the poultry world, it is bound to come, and is coming in the 'Orpingtons,' and particularly in the Buff Orpington, although my personal regard is equally good for the Blacks. The 'Orpington' Club in this country will no doubt follow the lines laid down by the English Orpington Club, and adopt the standard of perfection as published by them. In my opinion, from three years' experience with the Orpingtons, the Orpington Club will have a simple walk over in pushing their favourites. In fact, the breed will push itself here as it has everywhere else, and become ere long the poultry craze."

Another breeder writes: "To-day the Orpingtons are the most popular breed in England, and are sure to win favour in the United States. They combine the weight of the Cochin with the laying qualities of the Leghorn. What better fowl can anyone wish? They have strong, rather short, white legs, free from feathers, and white skin. Some breeders object to the white legs and skin, but I find that some broiler plants are bleaching the yellow skin of Rocks, as consumers desire white. With Orpingtons there is no need of bleaching. The rich flesh colour showing through the skin gives them a fine appearance when dressed. The shape of the Orpington is long, broad, and deep; breast broad, deep, and full, much better breasted than the Rocks, broader and more plump, as they are a larger bird. They are active, but contented and easily yarded. Chicks are very hardy, maturing early, and have been known to commence laying at the age of four months."

The opinion of an American poultry judge, Mr. E. S. Comings, will complete these references, all going to show, as previously stated, that the Orpington fowl is not only a popular one, but is universally so. "Any new breed when first coming into public notice must submit to severe criticism, and if its merit is sufficient to withstand this it means much to the breeder and fancier of the breed. That the Buff Orpington is one whose claim must be recognised no judge of fowls can question, so when I say that in this new candidate we have one worthy of our attention it is because I believe it. The popular ideas among American breeders in favour of the yellow leg and yellow skin is against them. Yet when dressed the Buff Orpington must present an inviting appearance, for their good size and plump bodies are surely in their favour. Many times they have been compared to the Buff Plymouth Rock, many saying that there is not enough difference between these varieties to keep them apart, but if you will study them carefully you must admit that they are two distinct varieties, and of the two the better specimens of the Buff Orpington are to-day far ahead of the Buff Rock in finished appearance. Note the curved lines in the outline of the Orpington, note how nicely he stands in equal balance, see how symmetrical each line as it merges into these curves that always please the fancier. Note that beautiful transparent buff outer colour—and it seems a buff peculiarly its own. Combs are nicely balanced in the better specimens and seem far more symmetrical than those of the Buff Rock, and every breeder knows how hard it is to breed a symmetrical comb on any variety of the Rock family. Any novice can see that in the Orpington we have a variety that can be made either a fancier's or a commercial fowl. In this variety, as in all buffs, there are several things to avoid in mating to produce exhibition specimens, namely : White in hackle, white in wings and flights, black in wings or tail. Of the two colour evils choose the latter, as black represents a colour quality, and it is but a step from black to rich chestnut, and from chestnut to buff; while white will never breed out, and Nature's rule of reverting to parent stock will annoy you for many seasons. The Buff Orpington has come to stay. Let us give it a cordial reception. That it is already claiming the attention of some of our leading fanciers the show room this coming season will have many fine specimens to prove, and we can offer no better wish than that in numbers and quality they may stand side by side the older varieties. Chicago, the Madison Square of the West, invites the fanciers of the Orpington to be well represented in the next exhibition. This will increase the admiration of this variety, placing them side by side those already recognised."

In concluding the rise and progress of this comparatively new breed in the States, and confirmatory of the writer's prediction of a few years ago, it may be mentioned that at the World's Fair held at St. Louis last year over 300 Orpingtons were exhibited, 152 of these being Buff, a much larger display than ever made at any Australian show, and exceeded by very few in England. That they will ever approach the Plymouth Rocks in the State is not even remotely suggested, but the already large army of patrons bids fair for the future prospects of the breed. The concluding sentences on the breed by an American poultry journal at the St. Louis fair was that it was a great day for the Orpingtons, and from all written and said about them above there is every reasonable hope of America becoming a great country of Orpingtons.

CHAPTER XXV.

Orpingtons for Meat and Eggs.

HAVING now given an outline of the circumstances which led to the inception of this breed of fowls, and an exhaustive history of its earlier troubles and ultimate triumph amongst poultry-keepers in every part of the world, it now remains to briefly give the attributes which were responsible for its present universality amongst poultry-men.

In England, it now holds pride of place amongst all breeds. Leghorns have many patrons, but fail in numbers in comparison with the breeders of Orpingtons. Plymouth Rocks have many devotees, but do not reach half the number of those who breed the English-made fowl; the runner-up in numbers being the Wyandotte. At the late dairy show in England, where the classes are for birds of the year only—no adults shown—the following are the numbers of exhibits of the principal breeds, which show the Orpington in England, as it is here, the most popular of all breeds, and most numerously exhibited.

There were on exhibition at the Dairy, Andalusians 27, Anconas 29, Brahmas 37, Langshans 37, Faverolles 44, Minorcas 60, Cochins 61, Hamburgs 64, Dorkings 77, Game 116, Plymouth Rocks 123, Leghorns 204, Wyandottes 343, Orpingtons topping the list with 350. Nor was this all, seeing that for the many new varieties of Wyandottes the extraordinary number of 18 classes had to be provided, which showed an average of 19 a class, while the Orpingtons had but 10 classes, being an average of 35 throughout. The largest display of Wyandottes was 32 Silver cockerels and 35 pullets, 35 White cockerels and 40 pullets, and 58 in the two Partridge classes. Coming to the Orpingtons, there were 31 Black cockerels and 45 pullets, 40 White cockerels and 65 pullets, while there were 59 Buff

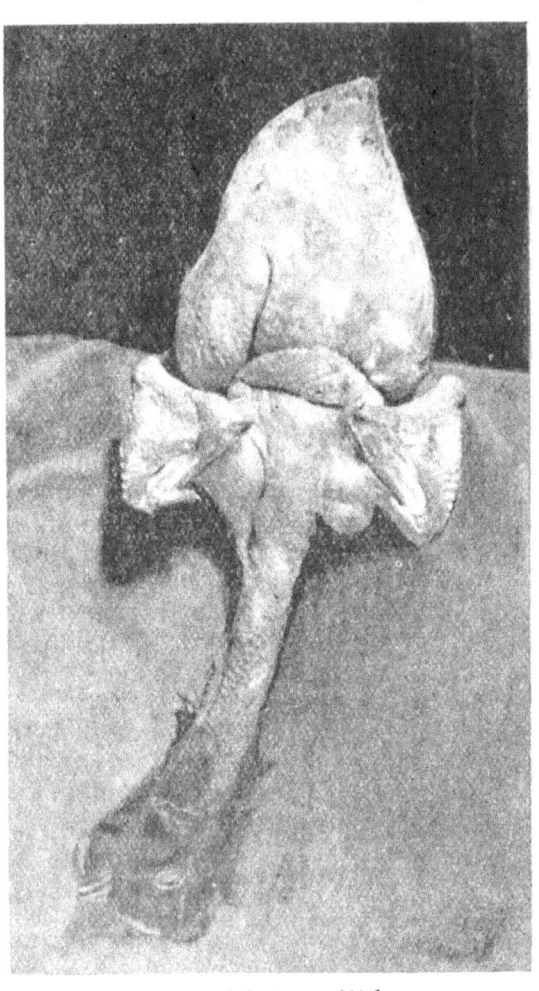

Well-meated Orpington Chicken.

cockerels and the extraordinary number of 70 pullets, showing that the latter colour are still the most found in England. What has brought about this popularity amongst the English fanciers is the simple fact, whether of the Black, Buff, White, or other colour, the birds are of big frames, sturdy growth, easily reared and managed, and, whether kept as egg-producers or carcases for the market, if there were such a thing as best breed, the Orpington fowl would be the one. Beginning with the Blacks, the market man will find them as quick growers as any fowl extant, and if well fed from hatching time, and otherwise wisely managed, the birds will reach 4 lb. each at 16 or 17 weeks old, the pure white skin and tender flesh warranting them a dish fit for any connoisseur in roasts.

Excepting the Blacks, all Orpingtons have white legs, and, should an export trade in poultry products ever become an established fact, there is not a doubt but Orpingtons of the various colours will form the larger bulk of the business. At every table poultry show in England, Orpingtons, principally Buffs and their crosses, have usually secured many chief places in the prize-list, the latest being at the British Dairy Farmers' Association's annual exhibition in October last; in large classes, containing all the breeds but Dorkings, Buff Orpingtons were third and v.h.c., while in a good class of pullets, Lord Windsor won second with Buffs, reserve going to the same breed.

In connection with exporting to London, the following from a London salesman, relative to a shipment of Sydney chickens a few years ago, may be repeated. The birds were largely Orpingtons and

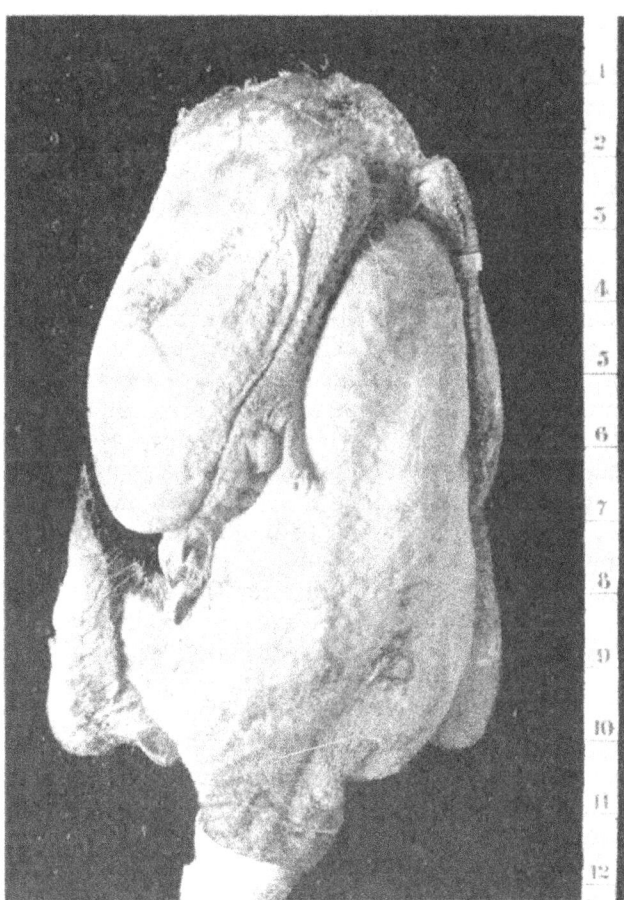

Black Orpington Pullet, showing a long, meaty breast.

their crosses, and were shipped through the Agricultural Department. "The chickens, ex 'Australasian,' made 4s. each, and were very fine. Only get them here earlier, and any quantity can be sold at from 4s. to 5s. each with no difficulty. They are the finest frozen chickens that come to our market, and the way they have been killed, dressed, and prepared, is deserving of every praise. Although there have been large quantities of Canadian, Russian, Hungarian, and other varieties, there is no comparison between them and the chickens from Australia.

The chickens referred to were shipped by Messrs. Boyd, of Gosford; Gray, of Paterson; and Hoffman, of Parramatta. They comprised a lot of good-sized birds, carrying plenty of meat, but certainly not the best that have been produced here. The "Australasian's" shipment left Sydney on the 5th May; the birds were sold in the first week of July, which is the tail end of the London season for frozen poultry. Considering that these birds, which could not be classed as the primest, realised so late in the season a price that will clear the breeders here over 6s. per pair, all doubts vanish about there being a profitable outlet in London for Colonial poultry of the right sort at the proper time of the year.

As market poultry, Orpingtons have short legs, free from feathers, wide and deep in body, full breast, the frame excellently suited whereon to quickly build meat; and for those who intend going in for breeding poultry for market purposes I can safely recommend this breed as one of the first they should try. In breeding them pure, and properly treated, they make most excellent carcases, and can be as cheaply brought to a marketable stage as any known variety, while for crossing purposes Mr. W. Cook's testimony will be conclusive, wherein he says:—"Cross-breeding in the past has not been looked upon favourably by old-school breeders, whose conservative notions have always hindered progress like this; but in many instances advanced spirits of their generation have indulged in this to a certain extent, and so many crosses have been tried with good results. If I may be permitted to give a piece of personal experience, I may say I· have learnt more of the real value of breeds through crossing than by any other means, and it was while crossing that I gained that insight into the characteristics of the various breeds which enabled me to choose out the best varieties with which to build up the various Orpingtons, which are now so popular."

Orpingtons can be bred profitably for the markets, while for those who prefer cross-breeding there are several breeds which can be judiciously used. A Dorking cock, if mated with eight or ten Black, Buff, or any other colour Orpington hens, will make a breeding pen of the very first order. They grow quickly, feather fast, and are in killing condition at almost any age from four months. A short-legged Colonial Game, or Old English Game, cock, if mated with the same number of hens, can also be recommended. The chickens from these will be more plump than the Dorking cross.

Coming to the breed's merits as egg-producers, such is of the very highest order, and despite the fact that it is an acknowledged principle that the best table qualities and an excess of egg-production cannot be found in any breed, Orpingtons go very near to dispel it; and, indeed, had the Buff variety equalled the Blacks at the various laying competitions in this and other States, such would have gone a long way in establishing the Orpington as the best all-round fowl, and the Buffs the best of the several varieties.

Regarding the laying competitions, there is no need to rehearse all the records made; suffice to say that as egg-producers the following figures are incontrovertible. ·At the second International Laying

Competition at the Hawkesbury College, which began on the 1st
April, 1903, and continued for three months, out of seventy pens
competing, one pen of six black Orpingtons entered the contest at
7½ months of age, and completed the year's work with 1,274 eggs, or
almost eighteen dozen for each hen.   The contest was both Interstate
and International, this pen of New South Wales Black Orpingtons
beating every breed and every pen, one excepted, that being
Wyandottes ; and, had weight of eggs been considered, the Orpingtons
would have won, seeing that the eggs weighed 25 oz. to the dozen,

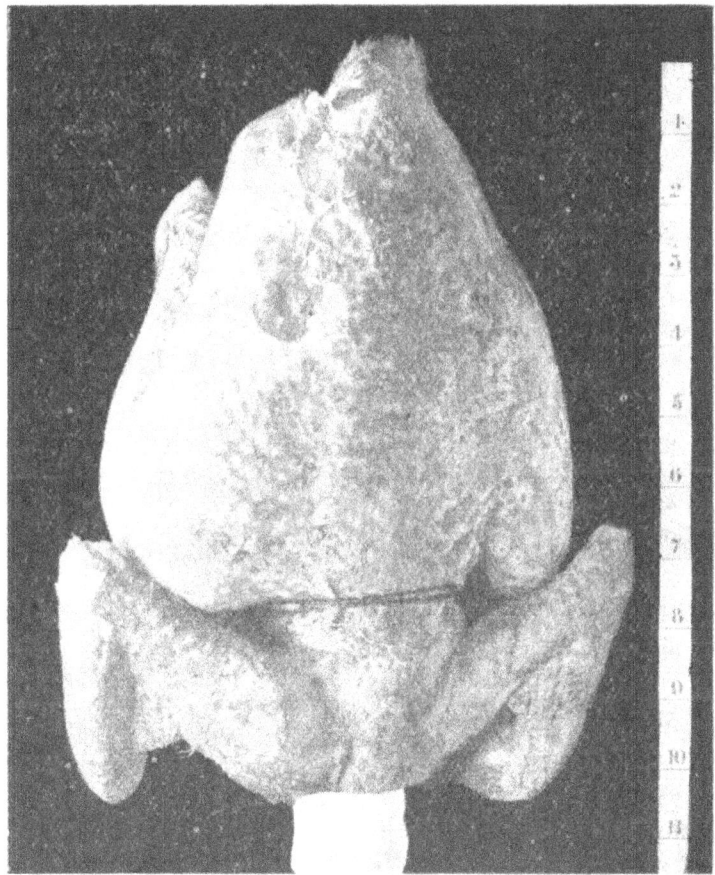

Buff Orpington—Faverolle's Cross 6 months old ; weight, 7 lb.

as against 24 oz. for the Wyandottes ; while, taking all the Black
Orpingtons in the competition, bad layers and good, 84 birds in all,
they averaged 168 eggs each, or fourteen dozen for each hen, a
performance of the highest order, and not responsible to any artificial
foods, spices, balanced rations, or other of the now many things
guaranteed to make hens lay ; the food was of the simplest, and those
electing to take up this breed, or any other for that matter, for the
purpose of a plentiful egg supply, need not go beyond the simple

formula which is embodied in Mr. Thompson's report, as follows :—
" The hens have been fed on the simplest diet possible throughout
the competition. The morning meal consisted of bran and pollard
mash at 7 o'clock. The mash was scalded with liver soup two days a
week, and on the other five days it was simply mixed with water, the
quantity given being an average of about one Imperial pint per pen,
the big eaters taking considerably over a pint, and the small eaters a
little under. In the afternoon, between 4 and 5 o'clock, the hens
were grain-fed, one pint, more or less according to appetite, of crushed
maize, and sometimes wheat. Cut-up liver was given twice a week,
at the rate of about 2 oz. per head. Shell grit was always before
them, and clean water was given every morning. In the way of green
food, rape was fed for three months during the winter, when the grass
was withered. For the other nine months, the only green food the
hens got was the natural grass in the pens. The rape was fed whole
in the leaf, at the rate of a dozen leaves to a pen every second day."

Coming to the later College competition, which commenced when
the preceding one closed on the 1st April, 1904, 100 pens competed,
and although there was a diminished egg yield in all the breeds from
the previous year, Black Orpingtons still held a high position, the 108
birds averaging 159·48, or over thirteen dozen for each hen. The
same number of Silver Wyandottes competed, these averaging 145·30
eggs, or a point over twelve dozen eggs each, and although the highest
pen of Orpingtons only got fifth place, they were only less than a
dozen each below the winner, while there were twelve pens of other
pens and varieties lower than the lowest of the Black Orpingtons. As
in the previous contest, the highest pen of Black Orpington's eggs
weighed 25 oz. to the dozen, as against 24 oz. laid by the winning
Wyandottes.

So far as the present 1905 contest has gone, the Orpingtons are
again placing beyond the region of doubt their reputation of egg-
producers of the highest order, Blacks again being ahead of the Buffs.
For the seven months beginning April of the present year, a pen of
Blacks have laid over ten dozen each at the Hawkesbury College ; while,
going to the Rockdale competition, of the fifty lots competing, two pens
of Black Orpingtons are leading with over twelve dozen each for the
seven months. The laying competitions in the other States exactly
confirm the experience here, for although but in rare instances have
a pen of Orpingtons topped the score, taking them as a breed in
every instance, they performed excellently. In the first Victorian
contest, which concluded on the 30th April last, a pen of six Black
Orpingtons from Wagga laid in the twelve months 1,228 eggs, or over
seventeen dozen for each hen, while all the Blacks in the same
competition performed just about as they did in this State, namely,
about fourteen dozen eggs for each hen, which goes to show that,
whether for eggs or meat, or both, of all the new breeds or old, or of
whatever inception or nationality, as a farmer's fowl nothing has yet
been introduced to this country from England or elsewhere to surpass
them.

CHAPTER XXVI.

Breeding Orpingtons.

WHEN writing on these fowls a number of years ago, I invited and received contributions from a number of its then prominent patrons, and were any testimony desired as to its profitableness, the very fact that in this age of new breeds and varieties the then advocates and breeders of Orpingtons continue doing so still, and what was then said by various writers has been verified by later experience. However, since that time other Orpington enthusiasts, and successful ones, too, have come on the scene, and divided honours with the old-time exhibitors to an extent that, when a leading show now takes place, the good Orpingtons are so numerous that many specimens of sufficient merit to win prizes in the olden days are now left cardless, and, although the bulk of the then prominent successful winners continue to win, at the same time a few of the later recruits are now disputing premiership with those of earlier experience, with the result that the breeder who can win a couple of prizes in the Orpington classes now-a-days is considered lucky indeed.

It need scarcely be said that to now secure show-pen honours, breeders must first secure well-bred stock, and give great thought to the mating, breeding, and rearing, and be thoroughly acquainted with pedigree, strain merits, and defects of the stock birds they use, and then, when all is done, it will be a good season indeed if two or three winners are produced.

The experience of Messrs. Ramsay, Pemell, Butcher, Grantham Farm, and others are already on record, and that of one or two of the newer patrons will now be given; nor has the success of these latter breeders been due to a lengthy purse, thus enabling them to import English prize-winners, but rather the success which has attended their breeding operations is further proof that just as we need not go to other countries for Orpingtons as egg-producers, neither is it necessary to go beyond our own shores for Orpingtons possessing that type and size which is considered essential in producing prize-winners.

Mr. E. Waldron, of North Sydney, is one of the most successful breeders in this State—a frequent prize-winner here and in Victoria,—and supplies hundreds of pounds' worth of stock to other States. Mr. Waldron's Orpingtons have kept him for years, and this is what he said to a representative of the *Sydney Daily Telegraph* :—

" I have been breeding for utility," he says, " for the past ten years, and have kept Black Orpingtons only. I am so satisfied with the results that I have no intention of making a change. For breeding I select close-feathered hens with broad shoulders and good chests. These three points they must have to suit me. A hen should also carry as much of her body in front of her legs as possible. The moment you get a hen that carries a great part of her body behind her legs, she develops fat, makes a poor layer, and her eggs will not give 25 per cent. of chickens. I have bred some very fluffy Cochiny birds, but find that they all develop fat very early, and at twelve months look like very old hens. For laying, breeding, or table, I would not care to keep many of them. On account of the fat and extra fluff, they get credit in the show pen for being low set, and will knock out a close-feathered bird that is actually shorter on the leg. I have proved this with my own birds. I am not a believer in either short or long legs. I like to see a bird with legs in proportion to its body. It is just as easy to breed one as the other. If you want the fluffy type, all you have to do is use a fluffy rooster, and you will always get it; but if you want layers, my advice is choose close-feathered hens, with bodies carried well forward, with broad shoulders, and good full chests."

Another prominent poultry-breeder who, within the past few years has gone in for Black Orpingtons, and with unprecedented success in the time, is Mr. H. Cadell, of Wotonga, Epping. This breeder appropriated the *Daily Telegraph* cup for the most successful Orpington exhibitor at this year's Poultry Club Show, securing two firsts and champion and two seconds in Blacks, and three of the firsts in Buffs, a record hitherto rarely approached, and as showing how this success was attained, and as a guide to those not already in the know, Mr. Cadell contributes the following, entitled " How to Breed Prize-Winners."

" In mating Black Orpingtons, the colour question is not so acute as in their younger relations, the Buffs; but to obtain the beetle-green so essential in the variety, care must be used in choosing a male to see that he is green all over and down on to the soft feathers covering the thighs, and fluff should also be green; he must be short on leg, full and round in breast, and dark in eye, in fact, a black or bull eye is preferable to the standard ' black pupil and dark brown iris.' In a stock cock, back short, broad at saddle, tail full and flowing—if carried a bit high, do not discard an otherwise typical cock, as a bird showing this fault is usually the sire of very short-backed progeny,—a neat head, clean-cut comb, and, although he may not prove a show-pen champion, his stock, if mated as I suggest, will be. Two years ago I purchased a cockerel at the Royal that could only get commended, giving £10 10s. for him, the winner going for £7 7s. The fowl I bought was a wonderfully blocky, large-chested fellow, and in choosing mates for him I went for short-legged, roomy hens, black in eye; the cockerel failed here, a couple in the pen showing an ample cushion, to get broad backs into the cockerels, and also to keep the tails moderate in size, with abundance of side hangers, and full saddles. The balance of the hens were tighter in feather, neat heads, and very deep in front as well as behind, depth being needed in a hen to give ample room for the ' egg department,' as in a heavy milking cow. Each hen was moved to the pen after careful study and with a definite object, all the while keeping in view the pedigree of each hen. A record of over forty-eight firsts, two silver cups, and numerous other awards for stock from this pen in New South Wales, Victoria, Western Australia, Tasmania, and New Zealand, all won at leading shows, is a guarantee of their show quality, and a pen of six exhibition pullets have laid from being mated end of July to end of October, and not one broody, speaks for them as farmers' utility fowls. I would strongly urge buyers to give more attention to pedigree if wishing to buy to show; and even after all the almost hysterical stuff one reads of so-and-so's marvellous egg-laying strain, a little quiet inquiry will usually prick that bubble, and one finds so-and-so's bred-to-lay strain are the culls from a show fancier's yard. An experienced breeder, who has a few years' show-pen successes to back him, is always more reliable than one of mushroom growth, and even the latter is away ahead of the dealer. Do not expect champions and quite perfect specimens; the former are always in demand at tall figures, while the latter have not been seen yet, although some point-judging cranks have scored fowls as high as 98½ out of a possible 100 at Sydney shows.

"While calling the Black Orpington a grand farmer's fowl, and by farmer I include all dwellers on the land, I think in a couple of ways the younger variety of the Orpington, i.e., the Buff, has a pull over the Black. For eating purposes I place the Buff an easy first, while the colour of the stub feathers, and there are always a percentage of these that remain, does not disfigure the carcase like the Black ones. I have found they more readily fatten, and put on more breast meat; then, by the poulterer, the white leg is much preferred. As Winter layers of nice tinted eggs, they run away from the Blacks, and as all my surplus eggs go to the leading grocers of Sydney, where each lot are weighed, I have never had a word about the egg being under weight, while their tendency to become broody early I consider their greatest point. During the past season I have raised about 500 chickens, about 400 being hatched by hens, and out of all I have set, but five were Blacks, and to a farmer early sitters are valuable, as early hatching means Winter eggs the following year, as well as meaty saleable cockerels by Christmas. When the Buffs are not required as sitters, if put away first time found on the nest after dark, they soon come laying again. As mothers they are unequalled, and many hens lay with chicks three to four weeks old, and still brood the chicks. If I had to choose one variety for commercial purposes, I would go straight and keep Buff Orpingtons only.

"As exhibition fowls they are very hard to breed to the one even shade of rich buff all over; but, after five years' careful mating, I find a much greater percentage of the chickens are coming true to colour, less black and white in tail and flights, and less leggy. To raise Buffs for show, careful inbreeding is an absolute essential, and if buying for producing show birds, ware the yard that is always introducing fresh blood. Just watch the show-pen, and though such haphazard breeders may occasionally score, the scientific breeder will average better. For getting show birds, use a sound, even-coloured male, and mate him to close blood relations, and, if of good pedigree, you will not be disappointed.

"I hatch principally with hens, and, when convenient, put two or three hens down at same time, and, when hatched, give all the chicks to one, coop her snug and dry, and the other hens can go back to laying. I have an incubator and brooders—Cypher's, about the best—but you cannot beat the hen. I feed on dry food, plenty clean, cool water, shade and shelter of the trees, and kill all weaklings as early as possible—that is, directly found. As they get to four to six weeks, I feed soft food of a morning, boiled grain at midday, meat twice a week, and dry oats, maize, barley, or wheat at night."

With all that has been now said about this breed of fowls, it will be apparent that for the exhibitor whose desire is prizes and their contingencies, or the farmer whose object is the greatest quantity of eggs and meat, there is no breed of domestic poultry which can be recommended to have greater all-round properties as that now so universally known as the popular Orpington.

Chapter XXVII.

PLYMOUTH ROCKS.

The following introduction to the above breed was given in the *Agricultural Gazette* in 1887, and with a slight alteration will suffice for this contribution on the one-time well-boomed in this country Plymouth Rock fowl :—

" Varieties or breeds of poultry have, like some animals and flowers, on introduction, been the subject of a boom or craze, and of recent years none more so than the Plymouth Rock. The show-pen records of this variety, for numbers, half a dozen years ago exceeded that of any other breed; the reported laying properties, table qualities, and size, combined with a handsome appearance, stamped them as the best all-round fowls, and the colonies, always eager to emulate the mother country, rushed them to such an extent that for some years the value of the importations from England was much in excess of any other breed. Both fanciers and ordinary poultry-breeders got smitten with what was then called the 'Rock fever,' abandoning older breeds of tested utility in their favour.

" However, in two or three seasons, the crisis was reached, with the natural result that they rapidly declined in public favour, and now only occupy a third or fourth rate place with practical poultry-keepers. Good exhibition specimens are found in but a few fanciers' hands now, while the numbers exhibited at our shows are only about the fourth of what they were a few years ago, newer varieties superseding them."

The alterations are but slight, the popularity of half a dozen years ago can now be read as thirteen or fourteen years, while the reference to the numbers exhibited can now be said to not reach more than a tenth of what they did in the early nineties. At that time they made the records in the shows for the quantities exhibited, while now they are beaten by several breeds and varieties. The breed was then, in Australia, England, and elsewhere, considered one of the very best;

55823   G

and so far as the country of its origination is concerned—America—it still holds the leading position, whether as a fancier's fowl for the show-pen, or a commercial one by the farmers, as judged from a profitable point of view. The evidence on the latter grounds being that it is bred in America, by fanciers and farmers alike, in greater numbers than any other breed or variety. As an egg-producer, the many experiments made at the various United States Agricultural Stations have in some instances shown it as leading all other breeds, while as a roaster, boiler, or fowl for export to England, it is approached by no other breed. Who of the numerous claimants originated them is of no note, except to say that a Dr. Bennett did produce a bird from a Cochin and Malay to which he gave the above name, but as they produced offspring with red and other coloured feathers they soon died a natural death, the present blue-barred fowls being the result of experiments with the American Dominique, Java, and other breeds. The name is that of a tribe of American Indians, and Mr. T. F. McGrew, who recently wrote a history of the American breeds for the United States Department of Agriculture, says: "When first produced no other name was needed; they were simply Plymouth Rock fowls, and became well known under this title the world over. No other fowl has ever enjoyed equal popularity in this country; and and we presume they are better known, and at the same time less understood than any other fowl of minor reputation. More has been written about them than could be read in years, and there have been almost as many opinions and theories placed before us as there are writers. This has caused considerable confusion until the alarm sounded ascribing retrogression to the breed, when the attention of those best able to cope with the difficulty became attracted, and marked improvement soon followed. There seems to be no condition, surrounding, or climate unfavourable to the Plymouth Rocks. Their constitutional vigour appears to have no limit. Where any fowl can live they will prosper. They stand confinement, and when allowed freedom prove excellent foragers. They are prolific in yielding medium-sized brown eggs of the richest flavour. Under all conditions they will produce fully as many eggs as any thoroughbred fowl."

Prior to the origination of the Plymouth Rock, America had no breed which they could actually claim as their own, and as the great want in all the American markets was a compact fowl, having close-grained flesh and yellow skin, and averaging from 6 to 8 lb. when dressed, the experienced breeder, with an eye to utility, saw in this robust constitutioned cross-bred one eminently fitted to supply such requirements, for beside weight they had the additional quality of putting on flesh readily, and laying a large number of good-sized eggs. In addition to this they had a pleasing form and a general appearance that would impress one with the idea that they were a useful fowl. From the time they were acknowledged as a pure breed and known as Plymouth Rocks, they became most popular in the land of their inception, and are at the present day exhibited in greater numbers than any other breed, possibly Wyandottes excepted. As showing the estimate in which the Plymouth Rock is held in that great country, the following figures are those recorded at the St. Louis Exposition during October, 1904 :—

	Cocks.	Hens.	Cockerels.	Pullets.	Breeding Pens.
Barred Rocks...	50	57	59	92	25
White Rocks	56	69	67	70	38
Buff Rocks	59	70	87	110	29

In previous poultry articles I have mentioned that we in Australia generally follow in the wake of the English poultry-men. In other words, when a breed or variety of fowls gets plentiful and popular in England, there also comes a run on them here. This was the case

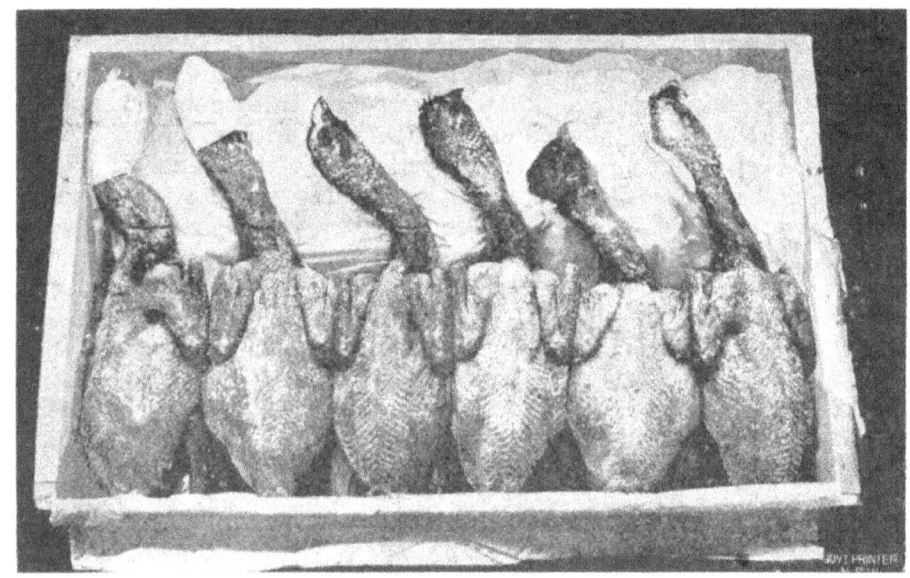

A case of 12 Chicago Plymouth Rock Chickens as opened in Sydney.   Papers removed from heads of four.

with Rocks, as in other breeds; but, strange to say, when Australia began to tire of the Rocks, such was not prompted by any like action in England, for the breed at the present time in that country if not bred for utility purposes is certainly still a favourable one with show-goers, as can be realised by the fact that at the Dairy Show held in the first week of October last, and which is for birds of the year only, there were 34 Barred cockerels and 40 pullets; 12 Buff cockerels and 10 pullets; and 23 Whites, or over 100 entries. While coming to the Crystal Palace and the International Shows, held a month later than the above, there was a combined entry of considerably over 300, much less than either Orpingtons or Wyandottes, but numbers sufficient to show that the Plymouth Rock holds a forward place in the English poultry world.

Regarding the merits of the breed, they are many; but the very fact of them declining in favour here is for the simple reason that other and newer sorts have greater economic merits. The Mediterranean breeds, the Orpingtons, and Wyandottes are more profitable fowls to keep, and this is the basis by which farm stock at the present day are valued.

Concerning the merits of the Rocks as table fowls, they are highly valued by the Americans, both for the family table as well as for the hotel trade, and although their English export business is a small one in comparison with that done by continental countries, still the many thousand head which go to London are chiefly Plymouth Rocks, despite the fact that a white-fleshed fowl is more favourably regarded in England.

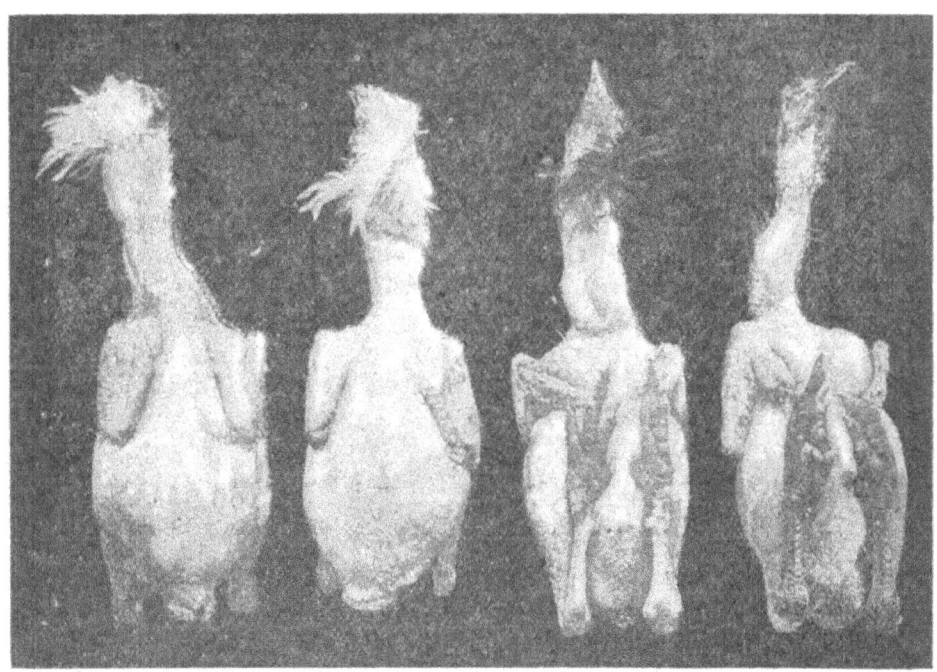

Four Chicago Chickens taken from case, showing fatted back and well-fleshed breast.

The late Commercial Agent, Mr. Lance, at the instance of the Director of Agriculture, secured during the past year in London a case of chickens from Chicago and forwarded them on here for inspection. All were Plymouth Rocks, Barred, Buff, and White, the different colours being recognisable in the accompanying illustrations. The birds were so extremely fat as to at once show they had undergone a course of fattening, and were approximately about five months old. The individual weights of the birds were as follows :—3 lb. 15 oz. ; 3 lb. 9 oz.; 3 lb. 8 oz.; 3 lb. 14 oz.; 3 lb. 9 oz.; 3 lb. 10 oz.; 3 lb. 12 oz.; 4 lb. 3 oz.; 3 lb. 14 oz.; 3 lb. 10 oz.; 3 lb. 7 oz.; 3 lb. 9 oz.; or a total of 44 lb. 8 oz., being slightly over 3¾ lb. each, consequently allowing for feathers and blood, the chickens would have been about 4¼ lb. each

prior to killing. They were purchased in London at 3s. 5d. each, so that allowing the moderate charge of 9d. for dressing, freezing, case, freight, insurance, landing and selling charges, would thus leave but 2s. 8d. for the exporter, who, in his turn, must have a profit, all showing that the American grower must get much less for his chickens than does the Australian breeder locally for the greater portion of the year for those of quality much inferior.

CHAPTER XXVIII.

Plymouth Rocks as Layers.

There is scarcely a doubt but what contributes to the continued popularity of Orpingtons and Wyandottes in Australia is the fact, as previously shown, of both breeds being good layers, some pens in the competitions having averaged over 200 eggs for each hen; while taking the good and bad together, they have shown a total for each fowl of from 12 to 14 dozens each in the twelve months.

Concerning the Rocks, although there are instances recorded in America where they made over 200 each, still in many cases they are considerably under this number. The American author I have already quoted gives the average yield as 150 eggs, and supplies the weight as 23 oz. to the dozen, a size which if accepted as a fair average in America certainly would not in Australia.

So far as the laying of the Rocks here is concerned, perhaps the worst that can be said of them in that respect is that not a single patron of the breed ever ventured to test it at any of the Hawkesbury College competitions.

At the 1904–5 test, which consisted of 100 pens of six birds each, seven pens of these were Americans, and one consisted of White Plymouth Rocks, owned by D. T. Roots, and accepting them as representing the American Rocks, their record is exactly that of the experienced breeder employed by the United States Government to write a history of the breed. At the close of the College contest the American Rocks occupied the 86th place in the hundred, the pen laying 775 eggs, or 129 for each hen. The eggs, however, averaged but 23½ oz. to the dozen; still to be strictly fair it should be mentioned that the eggs from a few pens of our own fowls weighed as low and some less, still there were over eighty lots whose produce weighed from 24 oz. to as high as 31 oz. for the twelve eggs. It should be said that age was in favour of the American hens laying larger eggs than ours, seeing that they were 10½ months old at the commencement of the test, whereas the bulk of ours were mere pullets of 7 and 7½ months old. In concluding this reference to the laying of the American Rocks, although it conforms with the bulk of the American writers, still one pen is not sufficient from which to draw definite conclusions.

In the 1905–6 Hawkesbury contest, not a single Rock appears, while for the 1906–7 competition, which begins on April 1 next, not one of the Plymouth Rock breeders have responded, these later experiences confirming my opinion as expressed in the *Gazette* in 1897 as follows :—"The majority of those who have given up the breed pronounce them as rather poor layers of eggs, which are small in proportion to the size of the fowls. The hens are much inclined to put on fat, which no doubt affects their laying properties. In spite, however, of their decadence, Plymouth Rocks have many good qualities. They are very hardy, stand damp, cold, and confinement well. They are good sitters and mothers, the chickens are easily reared, feather quickly and are not much given to disease. Those who wish to breed poultry for home consumption, and can afford to keep the chickens until seven or eight months old, will find the Rocks a good variety to keep. The cockerels and pullets of this age, if fed and otherwise well-cared for, will be very large and meaty, one bird alone being sufficient for a good family dinner ; but for the local or export market they cannot be highly recommended, for being fowls of a large frame, they are slower in developing than several other varieties, and, consequently, more expense is involved in bringing them to a marketable stage. During and for a few years after the Rock boom birds of this breed were to be found in moderate numbers in the poultry sale-yards of Sydney, but are now rarely seen there, Orpingtons and Wyandottes supplying their places."

White Rocks and Buff soon followed the origination of the Barred variety, and have also been bred and exhibited in this and other States, but to a limited extent, the evident reason being that more profitable sorts are available, and although at one time the Plymouth Rocks were fairly plentiful throughout the country districts, they have now disappeared, and are not likely to be resuscitated, or ever become a farmer's fowl.

CHAPTER XXIX.

LANGSHANS.

No other breed of fowls for a continuity of years has received such prominence through dispute and discussion as to type, purity, &c., as has that known as Langshans. The first arrivals of the breed in England were received by Major Croad, in 1872. They came from a district named Langshan in China. They were shortly afterwards exhibited, and the general impression then was they so much resembled Cochins that the writers of the day described them so, and the acrimonious discussions which have since arisen have been on this subject, and as a means to an end and finality in the dispute, English Langshan breeders have for a lengthened series of years been breeding them much more lengthy in limb than the Cochin, until the present day when the modern or exhibition Langshan has the distinction of being the tallest of domestic fowls; this lankiness or reach being that of limbs only, for the Langshan as at present known, although of apparent greater size, is actually no heavier than the short-legged bird of the early eighties. The breeders of this modern type press their claims for this legginess and sparse feathering on the grounds layers. We do not remember hearing of a single complaint of bad laying results from the White Orpingtons; and this is most encouraging, as we have aimed at producing birds of utility rather than a fancy breed, which in the future will indeed be the leading feature of all breeds, which are preserved to posterity. Leghorns have, of course, been looked upon as the best layers, for a great number of years, but Orpingtons have come in and taken their place very largely, and in the future are likely to do so increasingly, as the White Orpingtons have surpassed them generally, because, whilst possessing splendid laying results, they are also excellent table birds. This, of course, makes the Orpingtons more valuable. The great difficulty in connection with the small breeds, like Leghorns, consists in the fact that there are always a large number of cocks to be sold, or killed for table. These being small birds, of course, realise less than the White Orpingtons, which are good layers, and as fine table birds as they are layers. Then, again, this rule holds good where crosses are kept, for the cross-breds are finer and just as prolific layers as well.''

So far as White Orpingtons in this State are concerned, they have made but little headway. Four or five years ago Mr. J. E. Pemell, of Randwick, Mrs. Ewing, of North Sydney, and Mr. J. McComb, of Manly, exhibited a few at several of the Sydney shows, but speaking generally, they have not prospered, and are now rarely seen. At the same time they are becoming most plentiful in England, and have been exhibited in large numbers during the past year at some of the best shows. Some importations of this colour lately arrived in South Australia, from Mr. Cook's yards, and from correspondence I have seen in connection with them the birds appear to be superior in size to those exhibited here; however, seeing that we have White Leghorns and White Wyandottes in abundance, it is scarcely likely that Orpingtons of this colour will ever become favourites here.

Chapter XX.

Buff Orpingtons.

SEVEN years ago, when writing in the *Gazette* on Profitable Poultry-keeping, I mentioned that fashion had a good deal to do with the popularity of breeds or varieties of poultry. For a number of years one breed will take the public taste, then another will be in the ascendency; following this some old breed may be resuscitated, as in the case of the Old English Game at the present time, and displace some of the newer breeds. Then a particular colour of one or all the breeds may become fashionable, the bulk of the fanciers for a time patronising this fashion until something new appears. In the article in question I mentioned that the buff colour was the fashionable one, Buff Leghorns, Buff Langshans, Buff Wyandottes, Buff Rocks, and buff of other breeds, all having a call. William Cook was business man enough to take the tide at its flood by manufacturing a Buff of getting away as much as possible from the Cochin which the early specimens so much favoured; however, the short-shanked, clumsier type had many advocates, and from a few years after the Croad importations, through these differences, Langshan breeders have been divided into two camps, and hostile ones at that—the Croad advocates charging the other side with using Game largely to get the length of limb, the birds then by inference being cross-breds, while for the type favoured by themselves they are termed "pure Croads."

These breeders of the original type have of late years increased so largely that a club has been formed to preserve and encourage the breeding of the short-legged, big-bodied variety, with the result that there are now two standards for Langshans—the Croad type, the other one being that seen of late years at the Australian Shows, the "reachy" or as some English breeders term them "the Society type."

From 1878 classes were provided at a few English shows for Langshans, but for some years they made little headway, but once they became fairly well known in England they reached Australia. The first pair exhibited at any show in this State was at the New South Wales Poultry and Pigeon Society's Sixth Annual, held in the Temperance Hall, Pitt-street, Sydney, on 7th August, 1883. They were exhibited by Mr. J. W. Cumming, of the Sewage Camp, Bondi. The breed being hitherto unknown had no class provided, and were shown in the "any other variety class." In the following year, 1884, a class was provided at the same Society's show. Four entries appeared contributed by Mr. Cumming, and Mr. W. H. McKeown, of Gordon; the latter gentleman being a consistent breeder, importer, and exhibitor of Langshans until he retired from the fancy a few years ago. Mr. McKeown spent well nigh £200 in importations of this breed alone. As there were neither Orpingtons or Wyandottes at that period, and the Dorking and Spanish considered then as now, delicate, the Langshan appealed to both farmer and fancier alike as a good utility fowl, and in the space of a few years from the time the single exhibit appeared the numbers had reached up to 50, while coming up to 1895 and 1896 the exhibits ran up to 100, as many as 38 and 40 cockerels appearing in one class. This was about the record year, and from that period to the present day the numbers have dwindled down at every show in Australia, rarely more than a dozen appearing at the Sydney shows, while at many agricultural exhibitions throughout the country the breed has altogether disappeared. The large number of exhibits and exhibitors, the keen competition and excitement over the judging, and the big prices given for the winners, being, as far as this breed is concerned, a thing of remembrance only.

This decadence in public favour has to some been a subject of much wonderment, from the fact that the Langshans were undoubtedly good utility fowls, layers of large quantities of average size brown eggs, while as table poultry they were really excellent, big in frame, white in flesh, hardy and good thrivers; however, the patrons of the short-legged, flowing-tail, full-breasted Langshan of the eighties, who witnessed its evolution to the giraffe type of the present day, have no hesitation in saying that this changing in type very much affected the profitable qualities of the breed, and that realising this, fanciers and utility men alike dropped it in favour of newer breeds then appearing in the horizon of the fanciers' world, the Wyandotte and Orpingtons, length of limb in either breed being a show-pen evil, and considered detrimental to an all-round useful fowl.

In 1894, Mr. S. Gray, then sub-editor of the *Agricultural Gazette*, had a short article in the August number on this breed of fowls, the pair which illustrated it being drawn from photographs supplied by a then prominent breeder, Mrs. W. H. Webb, of Bathurst. The male bird won at New South Wales Society's Show in 1893, the hen being a winner of the previous year. The illustrations, which are reproduced, it will be seen, would do duty for third-rate Black Orpingtons of the present day, and is confirmatory of Cook's statement that he used Langshan blood largely in the manufacture of the Orpington. Indeed if such a bird as that illustrated was available at the present day, there are breeders in this State, who by selection and scientific mating, could in a very few years produce some of the modern Orpingtons.

As already shown, the blocky feathery type was the original Langshan ; at the same time, breeders here, as elsewhere, in order to get away from the Cochin type, did not object to a slight lengthening of the limb and shortening of the feather, and so long as the birds kept within reasonable bounds in this respect their popularity increased to an extent that at the time before mentioned Langshans were the most popular fowl in Australia. But as each succeeding year witnessed increased length of leg and reachiness, and those possessing this to the greatest extent being favoured by the judges, breeders called a halt, with the result that the one-time plentiful and profitable feather-legged black fowl is neglected to an extent that less than half a dozen breeders now patronise the show-pen with this over-much modernised breed of fowls. So far in this article I have confined myself to the ups and downs of the Langshan in the show-pen ; however, that does not tell all about a breed of fowls. The original Cochin-looking importations, and even later considerably modified arrivals, were all good layers and excellent table fowls, carrying plenty of white meat ; and these merits soon got talked and written about, with the result that a large bulk of the suburban poultry farms stocked Langshans, and a good many farmers did likewise, so much so that ten or fifteen years ago the poultry saleyards had a large proportion of these bulky fowls offering at their weekly sales, the egg market also showing evidence of the Asiatic blood, there being then a larger percentage of the chocolate-coloured eggs offering than appears at the present time. However, despite the decadence of the breed as an exhibition fowl, and the abandonment of it by the utility breeders, quite a number of the early patrons of the Langshans continue it as their only breed, or if more than one breed is kept, Langshans constitute one of the kinds. Those referred to are largely the old style Langshans, and although their now patrons would not think of exhibiting them with the expectation of winning prizes, they have great faith in their laying qualities, and, unlike the Plymouth Rock breeders, have ventured them in almost every laying competition, and with results eminently satisfactory.

At the Hawkesbury College 1903 competition, commencing in April of that year, seventy pens competing, two of these consisted of Langshans, and owned by Messrs. W. H. Ponton, of Tuggerah Lakes, and E. J. Winton, of Campbelltown. The six birds of the former owner laid in the twelve months 1,195 eggs, or within five of 200 each hen for the year. This pen came in sixth in the competition, beating sixty-four lots, including every breed and variety, and were within two eggs each of the well-boomed Mrs. Hansel's American Leghorns; while as an effective set-off to this trifling shortage, the Langshans' eggs weighed 26½ oz. to the dozen, while those from the American birds scaled only 24 oz., thus showing that the Langshans produced two or three pounds' weight more per hen than did the Leghorns, which produced a few more in number. Still, one pen of fowls amongst seventy, no matter how good a performance, is not a correct way to test production, and this was evidenced by the other competing pen of Langshans, they finishing in the forty-third place, the six birds laying 902 eggs; and here again the eggs were large, weighing 26 oz. to the dozen. The average of the two pens of Langshans were within a fraction of 175 eggs for each hen, considerably above either the White, Silver, or Gold Wyandottes, and the Orpingtons as well. However, it does not do to run away with some perhaps chance records on which to base assertions; and from the inception of these competitions I warned many to be specially careful about using a single test or individual pen to prove anything, and every competition held since the first all emphasise that there is no breed of fowls which can be safely termed the best layers. Strain is the principal feature, as all tests have shown; and while this may be questioned by a few as a determining point in egg production, every competition has overwhelmingly proved that strain governs production. Instances, of course, can be quoted where certain breeds did well in one test, and failed in the following. Such, however, is most easily explained. Very few breeders in Australia keep their strains intact—one illustration will suffice for many: A breeder whose fowls occupied a very high place in one of the competitions, the birds being of very mediocre appearance, so far as representing the breed was concerned, marking and type most indifferent, immediately made importations to improve the appearance of his stock. Such was the effect, but these new birds were not built for winning prizes at laying competitions, but were the more handsome sort to win prizes in the show-pen. The result is that the progeny have never since occupied a forward place; and, worse still, many of the progeny have gone to every State in the Commonwealth, being purchased and advertised as Mr. ——— prize-laying strain, and already it has been noticed their performances have been most disappointing. The remarkable and unfortunate feature of the incident is the fact that the breeder was unaware that the new birds would affect the laying of this strain, and advertised the introduction of the new blood. Those who have good laying strains of any pure breed of fowls, no matter how far removed from exhibition specimens, should hesitate before introducing new blood to improve appearance, except such be closely related to their own, while those who have exhibition birds of a breed, and which are good layers as well, are doubly blest; and that there are such the numerous laying competitions have shown. Reverting to the 1904–5 competitions, at which test all the breeds laid considerably less than at the previous one, a pen of Langshans secured thirty-third place in the hundred, the eggs again scaling the good weight of 26 oz. to the dozen. The pen of six laid 980 eggs in the twelve months; and again confirmatory of strain, another pen of this breed laid but 702, being within three of the foot of the list, but as a set-off to this the eggs weighed 27½ oz. to the dozen—an extraordinary weight for eggs from any breed of fowls. Coming to the present competition, which began on April 1, 1905, at time of writing the figures are just available to 31st of December, nine months of the test have expired. Of the 100 pens competing, two are Langshans, and both occupy advanced places; one pen (D. Frazer) has laid in the nine months 984 eggs, and occupies the third place in the 100 lots completing. The above is 164 eggs for each hen, and should they not produce another egg in the three months they have made a good record. The second pen (W. H. Ponton) has laid 894 eggs in the same time, which with what will follow in the balance of the twelve months will be further evidence, were such desired, of the productiveness of the one-time favoured Langshan; nor are the records shown confined to this State, they being still higher in Victoria, for at the first Dookie competition Ponton's Langshans laid over 200 eggs each for the year, and for a considerable time only five hens were competing, which were averaged among the six. Concerning this strain it may be mentioned that the foundation was laid from a setting of eggs purchased ten years ago from imported stock. The birds were then up to the standard, but would have no chance in present-day competition. Mr. Ponton's birds were not only from a good laying strain, but he has consistently kept them up to the laying standard by scientific mating, and excluding foreign blood. Then, again, at the Rockdale competition, Mr. E. J. Winton's pen of six Langshans for the nine months laid 1,038 eggs, being 173 for each hen, a truly wonderful performance, and all showing that Langshans, particularly of the old type, were and are good layers, and that those whose object is big brown eggs, and plenty of them, will not be disappointed by taking up a tested strain of this breed. As table fowls I cannot do better than reproduce what I said of them in 1898. Langshans are usually described as good all-round fowls. They are now bred long in leg, tight in feather, sparsely feathered on the legs, neat combs, and moderate tail, black legs, with white flesh, and layers of good-sized brown eggs. They are hardy, good sitters and mothers, do well in confinement, but either for local market or export are rather slow in developing, thus debarring them from first place for either purpose. Being birds of large frame, the chickens are rather bare of breast meat. The full benefit cannot be had from them until seven or eight months old, and although they are then like young turkeys, it would be questionable whether the price obtainable would pay for the thirty weeks' feed; but, as a fowl for home consumption, and killed at the age mentioned, they cannot be excelled, while for improving the table qualities of Minorcas they have many advocates, but never having had any experience of the cross, I cannot give results.

Mr. J. J. McCue, late poultry expert at the Hawkesbury College, in an article in the *Gazette*, said that Langshans and Minorcas were two good breeds to cross, the progeny being good layers and market birds.

At the English Dairy Show some years ago, when prizes were given for weight rather than quality, Dorking–Langshans won first, and Game–Dorking second. A few enthusiasts, however, practically tested the birds when dead, and found that the Langshan cross lost by drawing and trussing 20 oz., or a fourth of its entire weight; the Game–Dorking cross lost but 15 oz. in offal, thus proving that the Langshan cross was more apparent than real; and in a report on the subject one authority states that we may get large, hardy, and useful chickens for the family from a Langshan cross, but it would not produce a first-class table fowl for the market. In spite of this I am of opinion that they can be usefully employed in the improvement of the poultry of the farm yard, but I prefer those of shorter build than the present exhibition specimens.

Chapter XXX.

DORKINGS AND HOUDANS.

FROM the very earliest remembrance of all poultry men, Dorkings have been known, spoken and written of, first as an English fowl, and as the basis for the table poultry. Dorkings have been known in Australia from the earliest of our poultry shows, and many importations from England have taken place within the past twenty years, and a number earlier than that. A remarkable circumstance, however, obtains about the breed. No matter how good the specimens imported, the progeny rarely reaches the high standard of the imported parents, and it is most rare to find a breeder who takes up Dorkings to continue long with them. The chickens, it is said, are difficult to rear, and from all the sources from which information on the breed in this country can be obtained, the evidence goes to show that they are unprofitable fowls to keep. One thing is certain: years ago they were fairly plentiful in Australia, and in Sydney in particular, when as many as forty-five or fifty exhibits have been on view at one show, while now the numbers have dwindled down to half-a-dozen. Why this is so, few can tell. If they could be reared in sufficient numbers fourteen years ago to make a big display at the show, the climate has not so changed as to affect the rearing of them. The only explanation, and a reasonable one too, is that, even were the question of delicacy not involved, the birds are unprofitable, and this is the chief reason now-a-days for keeping fowls. The hens certainly are bad layers, and if kept for utility purposes would give a very poor return, while if for market purposes, no matter how prized the breed is for the table, other sorts will show a larger profit. The cocks are certainly to be recommended for crossing purposes, but so few of the breed are now reared, that were there a demand for a dozen good birds, such could scarcely be procured ; hence, as a farmer's fowl, it is unlikely they will ever have much call in this country.

Houdans have been frequently called the French Dorking, and always have the reputation of being excellent layers. However, whatever claim the breed has to such distinction in its native land, the English-bred Houdan, as a layer, is worthless, while the massive crests may be ornamental, but certainly not useful. Many show specimens have been imported to this State, but, like their English white-legged, five-toed compeers, have not been a success. Breeders do not take to them, and possibly for the excellent reason that other breeds are more suitable for their purpose.

Chapter XXXI.

GAME.

For a number of years Game occupied a leading position as an exhibition fowl at almost every show in Australia. Indian Game, British Game, and Colonial Game, a dozen years ago, had all large classes, keen competition ensuing. Of late years they have gone the way of a number of other breeds, and have but few patrons. Game fowls, of whatever sort, are proverbially poor layers ; this reputation, no doubt, being responsible for the non-appearance of the breed at any of the laying competitions. All the varieties have excellent table qualities, but the egg handicap is evidently responsible for the way in which they are now neglected. Game chickens, if hatched with and running with a flock of other breeds, with the same food and attention, will at any stage of their growth be covered with the desired breast meat, while the others may be comparatively thin of flesh. Game cocks, whether English, Indian, or Australian, if mated with Orpington or Wyandotte hens, will produce table fowls of the first quality ; and already such crosses have reached the London market from this State, and were favourably commented on there. Farmers, however, as a rule, do not care for this experimenting, and perhaps, after all, those that confine themselves to one breed, and make the best of it, can show more profitable returns than do those that keep a number of breeds and crosses. Still, despite the acknowledged poor laying of the Game, the Hawkesbury farmers, orchardists, and others in that wide district, breed the Australian Game largely, and at every auction sale-day in Sydney large numbers of these Hawkesbury chickens are on sale, and usually fetch from one to two shillings more than other breeds. In 1898 I contributed to the *Gazette* an article on Game, as follows, the opinions then expressed still obtaining :—" While the Cornish miners of some thirty years ago were building up a fighting Game cock to take the place of the Old English, and from which evolved the present Indian Game, a remarkable coincidence is the fact that about the same

time exactly the same process was going on in the Hawkesbury district
of this Colony, the breeds used by the old cockers for the required
purpose being almost identical with those used by the Cornishmen.
Both parties were working for the common end, namely, good fighting
birds, little thinking that a fancier's or show fowl would be the
ultimate result of their labours.   One great difference between the
Hawkesbury and Cornish evolutions is that regarding colour; the
birds produced by the miners is a new colour to the fancy, while
the Hawkesbury men followed strictly the line of the British Game,
producing their favourites in Black-reds, Duckwing, Piles, Brown-
reds, Whites, and Blacks.   So far as the general build of the birds is
concerned, a few years ago the Colonial or Australian Game and those
known as Indians were very much alike, large-bodied, strong in bone,
hard and close in feather, and carrying a great amount of flesh in the
breast; the breed has become a very popular one in this Colony, over
one hundred pens frequently appearing at the Sydney shows.   Of late
yaers, however, the craze for breeding long legs and giraffe necks on
Game fowls has been adopted by fanciers of Australian Game, short-
limbed specimens having now no chance for show-pen honours, with
a natural but sorrowful result of a decline in popularity, and what for
its many grand qualities promised to supply a want of a large well-
fleshed Game fowl as a foundation for table poultry, has for this cause
alone received a set-back in favour of the imported Cornishers with
not a single superior quality.   Australian Game are now bred to a great
size, 10½ lb. and 11 lb. being no unusual weight for cocks, the hens
going to 8 lb. or more.   As table fowls they are really excellent in
every particular, but like all other breeds that excel in this quality are
not prolific layers.   Of the many varieties we have, perhaps there is
none better fitted to breed pure for either local or export trade.   The
chickens, like all Game, are always in killing condition, and at sixteen
to twenty weeks are well suited for either the Sydney or London
markets.   Colonial Game I consider in every way fitted to supply the
market with table poultry, and this can be done either by breeding
pure, by crossing with other varieties, or by using them to improve
the ordinary farm-yard poultry of the Colony.   Any of the various
colours will do, preference being given to the pure blacks, they being
more, as the cockers left them, short-necked, short-limbed, wide-
shouldered, big-bodied birds, and of great hardiness."

Chapter XXXII.

THE MEDITERRANEAN BREEDS.

It is now a great number of years since the Mediterranean breeds were
introduced to Australia, and, excepting the Spanish, they have ever
been and continue plentiful and popular.   The shows, both in city and
country, whenever and wherever held, can always be depended on to
have a good display, particularly of Leghorns and Minorcas; while for
the poultry-farmer, whose principal object is eggs, the two varieties
mentioned are the most largely kept.   Indeed, while a flock of three,

four, or five hundred Orpingtons or Wyandottes are rarely if ever seen,
it is nothing unusual to witness such numbers of Minorcas or Leghorns,
and of the latter, particularly Whites. All the Mediterranean breeds
are good layers, still, as with other sorts, when they came to be
genuinely tested at the laying competitions, some of them performed
but moderately, while others have not only occupied premier positions,
but made records as well. At the Dookio (Victoria) College competition, which terminated last year, the winning pen was White Leghorns,
these making, up to that time, the highest record at any test by any
breed. The six hens laid 1,313 eggs, just on 219 eggs for each hen.
The pen which finished next was also of the Mediterranean breed, the
prolific Minorca, the six hens producing 1,228 eggs, or slightly under
205 for each fowl. The fourth prize were again Minorcas, with 202
for each hen ; the fifth and sixth places were also occupied by Leghorns ; the ninth pen were Andalusians, the six birds producing 1,159
eggs for the year, a number which would have won at some of the
other laying tests. However, there is little need to bring evidence as
above relative to the laying of the non-sitting breeds, all poultrymen
being aware of their prolificacy. Still, as with all other fowls, there
are poor performing strains, and at the test mentioned one lot of
Brown Leghorns occupied the lowest place with 636 eggs, or 106 for
each hen. However, an effective set-off in favour of the breed as a
whole is the fact that of all the Leghorns competing, except the above
lot, none laid less than 173 eggs each, a record unapproached by any
other breed. Coming to the present Hawkesbury College competition,
ten months of which have expired at time of writing, a pen of White
Leghorns have laid 1,239 eggs, being 206·3 for each hen, a number
almost equal to the best twelve months' performance, and which by
the end of the test will no doubt establish that pen at least as the best
egg-producers in Australia. Nor is the good laying confined to this
lot, another pen having laid 1,077, and another one reaching over the
thousand in the ten months. These, of course, are the highest numbers
made in the above time, the lowest being 574 ; and, as showing that
the egg production is governed in America by strain, as well as here,
one pen of Rose-comb Brown Leghorns have laid 1,027, and another
lot 740, from the 1st of April to the end of January. Coming to the
Rockdale competition, the ten months' laying is still better, for while
Orpingtons are at top, the second place is filled by White Leghorns,
with a record for the above period. The six hens produced 1,245, or
207·3 for each hen. As table fowls, Leghorns do not appeal to those
who make this branch a feature of the industry. At the same time,
poultry farmers who breed largely for eggs, in order to get laying pullets,
must of necessity hatch and rear a similar number of cockerels,
and these, of course, have to go to the market as table poultry, with the
result that any sale day hundreds of these Leghorn cockerels appear
there, and, as can be expected, being small and deficient in breast-meat,
do not realise the best prices. Of late years, White Leghorns are
being bred to a much greater size than formerly, and whether this
has injuriously affected the egg yield is a debatable question. The
effect is readily seen in the Sydney sale-rooms, the cockerels of this
breed being of much larger growth than obtained a dozen years ago.

Poultry-farmers and others will recollect how at some of the early laying tests American Leghorns became much talked about. They were of small size, had rose combs, and are moderately plentiful in some of the American States. To the staid or permanent poultry-breeder, the undue notice these birds got was as perplexing as unmerited. For the first few months they certainly laid well, but anyone with even a brief knowledge of fowls would never think of drawing laying comparisons from a three or four months' trial. However, the American birds got talked and written about, and before any definite results could be obtained here the country of their origination was exploited for laying strains, and within twelve months a number of these small American fowls reached here, and their reputed merits heralded throughout Australia. However, as is well known, three pens of American birds competed in the 1903 Hawkesbury test, with the result that the wonderful Hansel birds, the great American layers, finished, not on the top, as some had hoped, but in fifth place, the actual laying being 200·2 for each hen, being beaten by four pens of Australian-bred fowls, the winning Wyandottes laying 218, followed by Black Orpingtons, 212·2; third place was filled by Andalusians, which laid 207, and the fourth by Leghorns, with 204·1. The eggs of the American birds weighed but 24 oz. to the dozen, while the Orpingtons, Andalusians, and Leghorns, which beat them in quantity, also scored heavily in weight, running from 25 oz. up to 27 oz. a dozen. Even a pen of the now despised Langshan fowls laid within eight eggs of the American Leghorns, the eggs here, again, weighing 26½ oz. to the dozen. However, despite this, expensive importations continued from America.

Poultry-breeders anxiously awaited the result of the 1904–5 test, where seven pens of American birds competed, and here, again, the Australians beat them hollow, the American fowls finishing third, ninth, sixteenth, eighteenth, eightieth, eighty-sixth, and ninety-third in the hundred pens competing. The best record of the American birds was made by Rose-comb Leghorns, which was 193·3 eggs for each hen, weighing 23½ oz. to the dozen, Mrs. Hansel's, this time, only making 1,071 for the six birds, while the great H. Van Dresser's pen of White Leghorns were away down with 750, or 125 for each hen; in this instance, however, the eggs were 27½ oz. to the dozen. The above reference is not purposed to depreciate the American birds as layers, for, although not winning any of the tests, they occupied, as a whole, good positions, but rather to show that, while not doing any better or even so well as our own in laying, these American Leghorns have affected the breed here for ill in table qualities; for although, as previously stated, Leghorns are not bred for table poultry, yet the half of those reared have to be marketed for that purpose. These American Leghorns, both Whites and Browns, were of very small size. This was not only apparent in the importations which have been exhibited, but in the competing pens as well. In the College Expert's report he mentioned that the Hansel Leghorns weighed but 3¼ lb. on arrival, but in two months after they got up to 3¾ lb. each. Our own, or rather English Leghorns, go up to 5 lb. or even 5½ lb., and although some deprecate this great size, the Leghorn standard says "Large to be preferred, consistent with type." As I have shown, these American Leghorns for table purposes are of little account is now being realised in the proper quarter. The secretary of the Poultry Farmers' Co-operative Society lately called attention to this matter in the Press; but, more important still, in the country from where these small birds came, there is an agitation for greater size. Mr. J. K. Felch, a veteran poultry breeder, judge, and journalist, and holding a position in the American poultry world such as did the late Lewis Wright in England, in December last contributed the following to the American poultry Press. "The feed that makes muscle makes eggs. To use a dwarf because of prime colour is not good common sense in the poultry-yard. If by careful breeding we should raise the weight of White Leghorns to 5 lb. for pullets, 6 lb. for hens, 6½ lb. for cockerels, and 7½ lb. for cocks, they would become the most popular breed. This can be done and still preserve all the character and beauty of the breed. Take them off their Bantam legs, and raise them to a point of weight to make the males appreciated as early poultry meat, and make the breed far more popular and profitable."

It now having been shown that the large Leghorns lay as well or better than the small ones, and that as table fowls they are ever so much better, it follows that when new blood is wanted for either purpose—eggs or meat—we will, as heretofore, go to the world's breeding ground—England—for them, and no people on earth realise this to a greater extent than the cute Americans, who are the best customers English poultry-fanciers have, indeed for every fowl we import from England the Americans receive hundreds from the same country, and one or two of the Americans from whom the Australians imported are the largest English buyers, and publicly announce such in the poultry Press of their country. Fortunately the breeders here are beginning to realise that everything good appertaining to fowls need not be American, and to the writer's knowledge there have been some serious disappointments in the expected egg-production of these importations, the effect of such being that within the past twelve months there has been almost a cessation of these American poultry arrivals.

I have mentioned England as the world's breeding ground; the following extract from an article in the *World's Work*, entitled "Where Great Britain is Supreme," will be opportune. "Curiously enough there are a few elemental facts which have remained unchanged and undisturbed, and will remain so, through every legislative reform, and every proposal for tinkering with tariffs. One of them has a vital bearing upon agricultural problems, and it is the comforting truth that the British Islands have hitherto formed a great depôt for typical breeds of stock, in every market of the world. One instance of the creation of such a type stands out in the development of the English thoroughbred racehorse. In other forms of horses, in oxen, in sheep, in dogs, in poultry, this country has gradually produced the types which remain unbeaten by the utmost energy of the rest of the world, and this, although it is mainly private enterprise on our part against the State-aided and bounty-fed competition of our neighbours and rivals. But nothing would have availed us were it not for certain underlying natural advantages which we possess in the climate, the soil, and the geological formations of our island home. Short of a volcanic cataclysm, these advantages we shall retain. There are in fact more varieties in underlying soil and rocks in the United Kingdom than in any other thickly inhabitated area of the same size, and this is the key to the problem of breeding."

Reverting to the other colours of Leghorns—Buffs, Blacks, Duckwing, and Mottled (Anconas), all are good layers, the small size being a handicap to their recommendation as a farmers' fowl. Minorcas have made great laying in several of the tests in this and other States, and although there is no good laying strain of the breed at the present Hawkesbury test, at Rockdale one lot of six hens laid 201 eggs each in the ten months, and any one who wants, above all other things, eggs, will not be disappointed in Minorcas. As table fowls they are not the best, still on the breaking up of one or two poultry farms this year, several hundreds of this breed were purchased for export, as boiling fowls, and when killed and dressed were attractive in appearance, big, fat, and white in skin. Andalusians are excellent layers of large white eggs. They were third at the 1903–4 Hawkesbury test, and made conclusive records at other competitions, and although not so plentifully bred as either Leghorns or Minorcas, those who patronise them would not give them up in favour of any other breed or variety of fowls. Spanish complete the Mediterranean breeds, this one-time useful fowl being now almost out of existence, and not likely to be resuscitated for any purpose.

CHAPTER XXXIII.

FAVEROLLES.

THE above is the name given to a now recognised breed of fowls, the district of Faverolles, in France, being responsible for its name, in the same way as the bulk of farm-yard fowls of Sussex in England are named Sussex fowls; and in France the Faverolles answer to what the Surrey or Sussex birds do in London, *i.e.*, make the highest price in the Parisian market, and has been bred by a people whose sole object has been to make profit. The Faverolles is just like the bulk of what our other breeds are, or rather were, a cross-bred, produced from two or three other breeds, as were Orpingtons, Wyandottes, Plymouth Rocks, and others, with the object of producing a variety that would fill the bill as winter layers, and make big, early, white-skinned roasters.

M. Rouillier-Arnoult, Director of the French Poultry School at Gambois, says—"To obtain a true explanation of the breed, it is necessary to go back about forty years. The district at that time around Faverolles possessed only a common kind of poultry, together with some Houdans. This was the period when Cochins, Brahmas, and Dorkings first appeared in France, and the poultry-keepers crossed these or some of them with the common fowls, and also the Houdan. The progeny from these crosses had the size and body of the male

55823   H

parent, whilst retaining the delicacy of flesh which has made a favourite of the fowl from the Houdan district."

M. Rouillier-Arnoult then gives a description of the appearance of the birds, and finishes with—" This race is particularly recommended from an industrial point of view. The chickens of Faverolles are exceptionally hardy and easy to rear, a great advantage, which breeders in this country are not slow to appreciate."

This statement as to the origin is not disputed, for in an old English journal in the possession of the writer, printed twenty years ago, a correspondent, inquiring about Faverolles, was told that the fowls known by that name provided the greater portion of the poultry markets in the province of Ile-de-France, and that they were nothing else than the produce of a cross between the French Houdan and the Brahma, and that there was no occasion for him to send to France, as the progeny from the above two breeds would supply his wants. However, since the above date, both Buff Cochin and Dorking blood has been introduced, the former being responsible for the subdued buff or wheaten colour of the hens, the leg feathers, and the docile disposition of the breed. The Houdan gives them the characteristics of whiskers and beard. The Dorking is most noticeable in the male birds, as seen by the pinkish white legs, five toes, black breast, and distinctly Dorking comb. The Brahma and Cochin are also responsible for the brown eggs, and for the general carriage or type of the bird. Although breeding for feather was not studied by the original producers, yet the different and more or less systematic ways of producing these birds have brought into existence several colours ; but so far as the English show-pen is concerned, only the Salmon and Blacks are recognised. In France, the English salmon-coloured variety is called Faverolles samoun. There is also the Faverolles Brahma in France, marked similarly to the breed of the latter name. The Black owes its origin to the Langshan and the Faverolles-Concou.

Mr. T. R. Robinson of the South Eastern Agriculture College, Wye, Kent, lately wrote of this breed—" As to the question ' What are Faverolles ? ' one might reply, that as the pig in Ireland is said to be ' The jintleman that pays the rint,' so Faverolles are the fowls that bring in the cash to the more advanced commercial keeper of poultry in France. Having had some considerable experience both in England and France on this subject, I will give a few words of description first on the foreign bird, and then as she is made in England may not be out of place here. French Faverolles, as bred by practical men, are by no means true to colour; nor have they any other fixed characteristics. The producer wanted something easy to rear, quick to grow, and with great aptitude to fatten. These ordinary trade factors have existed many years, long before any Fancy element stepped in. It is easy to understand then that with many workers, and many buyers, that there should be great divergence of type, brought on, no doubt, by force of circumstances and greater or less judgment. Surely it speaks well for the people who, without method, with little knowledge of pedigree, and on a very mixed foundation stock of country-bred Houdans, evolved a bird, which on the market superseded all others; hence the variety of type and colour. That Faverolles contain the blood of Houdans, Brahmas, Concou de Malines, and, later still, Langshans, there can be little doubt. A few French breeders have classified them according to colour, yet the greater aim of the growers is not colours, or four-toes, or five-toes, whiskers, or no whiskers, but to breed a bird quick to grow and fatten. From an English point of view it would be absurd to say that they cannot be improved. They may be improved or spoilt. What in my opinion is wanted is quality with early maturity, with as many winter eggs thrown in as possible. Providing these objects are kept in view, there can be no harm in grading them to colour and uniformity of comb, legs (shanks) etc. By doing so they will become a breed that will probably appeal on its merits to a good many poultry raisers. It is not my object to 'write up' or 'write down' any variety of fowls. Personally I am a lover of many breeds, but to those who have the accommodation and ask the oft-repeated question ' What shall I keep ? ' the reply to some is ' Try Faverolles.' "

The late Mr. Harrison Weir, in his " Book of Poultry," says :— " Faverolles are a table fowl, and should be judged on the poulterer's bench, and there, featherless, all its qualities as a table fowl, rightly, carefully, and most scrupulously considered. The flesh, the texture, the quality, the fat, and the disposition of it, the skin, its thinness, tenderness or toughness, and the colour ; and with it all, the fowl should be properly proportioned, not with big thighs and legs, and long shanks, or small thin-made wings, or attenuated breasts, with a deep fleshless keel bone, but all should be kept in unison, square and meaty ; keep to this and it matters little what the feathering colours are. This appears to me to be the method adopted by the French, and for utility this is right. Having examined a large number of Faverolles at the Dead Poultry Show, I have come to the conclusion that it would not be so very difficult for the adept in such matters to evolve from the material, as now presented, such a utility breed as described. After careful inspection, I found that some of the very best were four-toed, yet, in some instances, a small toe only just indicating a fifth. This being so, perhaps it would be better to formulate the breed as one with only four toes, though my own experience has taught me that the most cloddy, thick-made, dunghill fowls have been those with five toes. Again, most of the best framed meaty fowls were those with clean shanks, and these very white. This being so, I would suggest that if the Faverolles is to become a farmyard breed, and if carrying the same shapely body, that it would be far more valuable as such with clean, clear, featherless shanks. Beyond this, I see no reason for any alteration in the Faverolles as it stands. It is a breed that should win as a prize bird when fatted and killed and shown as what it really is, and judged as a table fowl by what it is supposed to be."

Mr. Weir's suggestion as to the clean legs and four toes was duly discussed in England a few years ago. Some breeders were agreeable to the proposals, but as the fifth toe and feathered legs were more difficult to secure, the fanciers adopted such, and embodied it in the standard, with the result that, for show purposes, birds minus the fifth or Dorking toe are disqualified from receiving a prize, whilst more or less leg and foot feathers is also a show pen essential.

Mr. J. P. W. Marx, of Basford, Nottingham, an authority on the breed, contributed the following article to the late Mr. Lewis Wright's book :—

" Faverolles have for some time been common in the northern part of France, where they were regarded as simply useful fowls. They are the result of crosses to produce good layers, particularly in winter, whose chickens are strong, hardy, and quick-growing, with thin, white skin, and fine bone, abundantly covered with meat, and lending themselves readily, if need be, to artificial fattening. Brahmas or Cochins, Dorkings, and Houdans were used to produce Faverolles, and as the different varieties of those breeds were used indiscriminately, the Faverolles are met with of various colours, yet with well-defined characteristics of habit, shape, and quality. A few seem to have been kept in England about 1892 or 1893, but little was heard of them till 1896 ; since then they have become scattered all over the country. Whatever the colour of the Faverolles, the general characteristics are the same. In both sexes the comb is single, upright, medium in size, with neat serrations and free from coarseness. This is a difficult point, since of the breeds which were selected to make up the Faverolles, the Dorking alone has a single comb, which falls over in the hen. The peculiar combs of the Brahma and Houdan are strongly hereditary, and thus all kinds of combs crop up in the Faverolles, and most careful selection is required to get and retain the correct type. The beard and muffling should be very abundant, the beard thick and full rather than long and thin. These, again, being only found in one of the original breeds—the Houdan— are difficult to breed ; indeed, the head of the Faverolles is one of its most characteristic and important features. The head itself is broad and short, with small, thin wattles, and stout, short beak. The short, stout neck is thickly covered with rather close-fitting hackles. The body is broad, deep, and wide ; the back very broad and flat ; the breast is also broad, with the keel-bone deep and prominent ; the whole giving a sturdy, massive look to the fowl. Greater length of keel and back is seen in the hen. The wings show boldly in front, yet are distinctly small. The thighs are short and set wide apart, with the knees quite straight. The shanks are of medium length. The legs should be fairly stout in bone without being coarse, and be slightly feathered on the outside down to the end of the outer toe. The leg feather should be soft in texture, with no sign of the vulture-hock too frequently met with. The toes are five in number, and the extra or fifth toe, as in the Dorking, should be clear and distinct. The tail feathers and sickles are full and broad ; the sickles incline, however, to be short in length, and are carried rather upright, as in the Brahma ; a large tail with long sickles carried low or straight is not in keeping with the build of the bird. The tail of the hen is fan-shaped, and carried rather high. The colour of the Salmon Faverolles cock is quite different from that of the hen. Some are a mixture of black and

silvery white, like the Silver Dorking ; others, which have the preference, are warmer in colour, like the Dark Dorking.   In the exhibitions
salmon cock, the beak, legs, and feet are white : any pink colour on the
legs should be dealt with severely if it is too prevalent, and should be
eradicated.   The skin also is white and very fine ; a coarse red skin is
a distinct fault.   The face, lobes, and wattles are red, nearly concealed
by the muffling and beard, which is black ticked with white.   Neck
and saddle-hackles are straw colour, quite free from any stripe,
although many cocks still retain the Brahma hackle, and probably
will do so for some time yet.   The breast is black ; very few are sound
in breast colour ; the majority show white mottling, particularly
towards the bottom, others even have feathers tipped with bronze or
red.   More latitude is allowed with the back and shoulders, which
may be a mixture of black, white, and brown.   The wing-bow is straw-
colour, the wing bar black, and the outside of the secondaries white.
The tail, under colour, and thighs are black ; the tail coverts may be
brown.   Some cocks with much less black in them have the breasts
mottled with red and white, and the back and shoulders a rich red
brown ; these are very handsome, but not in accord with the present
standard.   The salmon hen is much like a wheaten game.   The head
and neck are a wheaten brown, broadly striped with a darker brown.
Beard and muffling (both are much heavier than in the cock) are
a creamy white.   Back, shoulders, and wings a wheaten brown,
the colour running lighter on the sides until it meets the cream
colour of the breast, thighs, and under-colour.   Primaries, secondaries,
and tail are wheaten brown ; these at present are very imperfect,
for a great deal of black, or white, or both is to be found in
most hens.   Face, wattles, legs, and feet are the same as in
the cocks.   The definition of the colour as " wheaten brown " is
not a happy one ; it may mean the warm brown of red wheat, or the
much lighter shade of white wheat, and the latter seems to be the
colour which is required.   The fashionable salmon hen is a warm
cream colour with a pale brown colour on her neck, back, and tail ; a
delicate pink or salmon shade in these colours is preferable to a faded,
washed-out whitish colour.   Any trace of buff, gold, or hard brassy
colour should be discarded.   There is a very handsome strain of what
may be called red wheaten brown hens ; the back and sides are
blotched with a deep chestnut brown, which runs on to the tail, and
the hackles are broadly striped with the same colour ; they have a
rough hardy look, but are too dark and red for the colouring of the
standard.

" The Ermine or White Faverolles are marked like Light Brahmas,
and, remembering their origin, it will be found quite as difficult to
obtain the clear, densely-striped hackles with pure white body colour,
free from ticking.   In mating Salmon Faverolles, comb, width of
back and between thighs should be attended to in both sexes.   The
comb should be free from side sprigs, and, if possible of fine quality in
the hen, and upright.   The best combs procurable should be used, for
faults here are sure to appear in the chickens.   A cock with heavy beard
and muffling is valuable as a breeder. His neck and saddle hackles should

be a yellow straw shade in preference to white for cockerel breeding; a slight stripe or ticking of brown or brownish-grey may be tolerated in a pullet breeder. Hens with any black in the hackle, even at the tip, should be cautiously bred from unless it is known their mother was better than they in hackle feather. The feather itself should be rather short, but broad, to give room for the darker centre. The breast of the cock should be solid black from throat to thigh. Many are ticked with white and a few have a mottling of red or brown, and these are likely to breed better coloured chickens than those ticked with white. The sounder the black of the thigh and under-colour the better. Cocks showing much white breed cockerels lighter than themselves and pullets too weak, almost white in under-colour. The tail coverts should be a dark chestnut brown in a pullet-breeding cock and the rest of the tail black. The sheen on the black throughout the cock should be a rich metallic bronze, not a beetle-green shade.

"The hens should be as near the Standard colour as can be obtained. The weak points are wings and tail where black and white are sure to be found. Hens with much white in wing should be mated with a bird sound in wing, with very little white on the ouside of the secondaries, plenty of bronze on the shoulders, and very little white ticking in his under-colour. The brown colour of the tail may be improved by selecting a cock with abundance of coppery-brown lustre and brown tail coverts. If the tails of his daughters show an improvement he may be mated up next year with the best of them in that respect. The shaft and down of the feather quite to the skin should be a creamy or wheaten brown. Hens with a brown or ashen-grey down throw a number of pullets with black in wing and tail. Faverolles chickens are very hardy and easy to rear, either artificially or naturally, providing that they are given as much liberty as possible, for after the first week they are keen foragers on their own account without being wild and prefer food of their own finding if it can be obtained.

"The framers of the Standard also appear to have drawn it up from a meat-producing point of view; perhaps rightly so, for the breed has found much favour with the fatters, some of whom declare it to be the nearest approach to the old Sussex breed they have met with for some time. The chickens are white when hatched, and their nest feathers are also white; with each subsequent growth more colour appears, but only in the final change do the cockerels acquire a solid black breast, so the weeding out process must not be too hastily conducted. They grow and mature very quickly until the final change into adult plumage, when, like Brahmas and Dorkings, the feathers come rather slowly. The claims of the Faverolles as a table fowl seem to have obscured its excellent laying qualities. Helped by its early maturing quality, however, the Faverolles is also a good winter and spring layer, not easily checked by climatic changes. The eggs vary in colour from white to deep brown—most usually they are a pale brown. Pullets' eggs are deficient in size, but those from mature birds are above the average. The hens are slow to come on broody, though exemplary sitters and mothers, and if checked from broodiness soon recommence laying."

Chapter XXXIV.

Faverolles in England.

As stated in the previous chapter by Mr. Marx, little was heard of the breed in England till 1896, but each year since that time they have increased amazingly at the shows, and to further encourage their advancement and popularity a club was formed, whose first duty was to draw up a standard by which the birds were to be judged, and such standard now being embodied in the English Poultry clubs' standard has had the effect of stimulating breeders, by the offering of good prizes at the various English shows, the numbers of the Faverolles at some of the important fixtures now approaching many of the older breeds.

As might be expected, the enterprising Americans did not allow this new breed to escape notice, and, shortly after they became plentiful in their adopted land, several of the American breeders visited both England and France, bringing from the former country quantities of the recognised show specimens, and from France all the colours of the breed that were to be found in the Faverolles district. Dr. Phelps, a noted ornithologist and poultry breeder, has been the largest American importer, and has spent thousands of dollars on the breed. Other enthusiasts have also spent much money in importations, Mr. J. F. Crangle, a well-known American authority, writing of them in an American paper, says: " The recent importation of Faverolles to this country by Mr. Joseph B. Thomas has brought into prominence this new French breed. The Faverolles have an advantage over others, in being prolific layers of good-sized eggs, which average 24 oz. to the dozen. The colour of the shell would be classed as pale, or very light brown. The fowls are large in size, heavy and plump, with very long full breasts, which carry considerable white meat. There are several types as well as colours. Those selected by Mr. Thomas are known in England as Salmon Faverolles. Having been formed by the union of Houdans, Dorkings, and Asiatics, they combine the good qualities of all these gained under the careful guidance of those who originated them, whose sole object in their production is for gain."

In England, the Faverolles now occupy a prominent position as one of the best utility breeds. All the fanciers journals in that country specially recommend them for both meat and eggs, but particularly the former, this being confirmed by the fact that at all the table poultry competitions it is the one largely shown and favourably commented on by the market poultrymen.

During the past year the *Fish and Poultry Trade Gazette* when writing up the Autumn poultry trade, and the breeds best suited for choice table fowls, said of Faverolles :—" A recent addition to British poultry comes to us from France, and is probably a mixture of the best French breeds and our own Dorking fowl. The cocks are rather similar to the Dorking in colour, but have a muff and beard on the head, feathers down the shanks, and *five* claws on the feet. The female is cream-coloured, except the neck and back, which are red-brown. They are capital table birds and easily fattened, and, as they are hardy and splendid layers in winter, are becoming one of the most popular breeds. They are being largely employed in Ireland by the Chamber of Agriculture for improving table qualities, and before long their distinctive head points will be seen largely in the English markets. In the autumn they should be bought with caution, as they are such an early laying breed and such quick growers that they are developed before most of the other heavy breeds, and become slightly hard. They are at their best about four months old, and even at an early age make splendid spring chickens."

In Ireland the County Council recommended them to the farmers and cottagers, and roadside rearers, on account of their hardiness, equal table qualities to the Dorking and good winter laying, and at several of the distributing stations in that country, where settings of pure bred eggs are sold at nominal costs to all applicants, to improve the poultry stock. A few of the best breeds only are kept. In County Longford there are ten of these stations each having thirty pure-bred fowls of one or more breeds, and from these stations from January of last year to the end of May, 1,239 settings of hen eggs were sold. At one of these stations—Ballyreaghan, 113 settings of Faverolles were sold, while at Cornakelly 148 settings of Faverolles were disposed of, a greater number than of any other breed, Orpingtons excepted. Nor do the Irish authorities overlook the good qualities of the Faverolles, for at even the smallest show in that country, classes are provided for them.

Chapter XXXV.

Faverolles as Table Fowls.

Many interesting experiments have been made of the growth of the Faverolles chickens in comparison with other breeds. One was carried out by Monsieur Grange, at the School of Poultry Culture, Gambois France, but as all the records are given in grammes and kilogrammes, the weights although heavy for the age, appeal to few here, and were of small importance beside the series of experiments made in England last year, by Mr. E. Brown, at the Theale College Poultry Farm, Reading. The experiments were the most exhaustive yet made in any country, and although Faverolles occupied a prominent place in all the series of tests, the appended detailed accounts are given here, to show the cost of producing chickens of any breed up to a marketable age and size in England, and although the conditions there in the way of poultry foods are not exactly like those obtaining here, still they afford a fairly accurate guide to Australian breeders, as to the cost of production of table poultry in England, and enable them to form an opinion as to whether, with the distance and other handicaps, we can profitably compete in the world's market with our poultry products.

The experiments commenced in March last year, and concluded on 5th July. The commencement here refers to the time the eggs were placed in the incubator, the hatching being completed on 4th April. There were four lots experimented with, consisting of thirty each of Faverolles, White Wyandottes, Buff Orpingtons, and cross breeds. The mortality was very slight throughout the period of observation, three chickens of the 120 dying during the experiments, and as showing how strictly the tables and figures were kept, the three losses were regarded as part of the cost of rearing.

The chickens were treated throughout in identically the same manner, and were hatched from eggs produced by stock kept on the College Poultry Farm. Each lot were hatched on the same day, in incubators of the same class, when dried off (twenty-four hours after hatching), each lot were accommodated in a brooder of the same class. For two weeks they were kept in heated brooders without grass runs, for two weeks longer in heated brooders with limited grass runs, and for one week longer in the brooders without heat. When five weeks old the cold brooders were removed, and the chickens were placed in a large house without perches, remaining there until the end of the full period of twelve weeks ; these houses were in large grass runs. During the whole time they were fed in identically the same manner, no attempt was made to force growth, and were treated in the natural way. For the first five weeks the brooders were kept in a paddock on the farm, and moved to fresh ground daily. Throughout the entire time careful attention was paid to cleanliness. The houses to which they were removed at the end of five weeks varied somewhat, but each contained about 234 cubic feet of air space. The runs contained about 280 superficial yards each, or about 9 rods, and were laid down in grass. They were well sheltered on the north by large chestnut trees, and were planted with fruit trees, but additional shelter was provided by means of hurdles. At the time the eggs were placed in the incubators, (13th March) their market value was slightly under a shilling a dozen, and in the following calculations they have been estimated at one penny each. The hatching averaged about 70 per cent., thus nearly forty-three eggs were required to produce thirty chickens at the time of hatching, and the egg cost of each chicken was 1·43d. The cost for oil burnt in a 100 or 120 egg incubator is about 3d. per week, and allowing four weeks for regulation and complete hatching, this gives a total of 12d.

Providing for infertiles taken out, and taking two lots in one machine, a total of 6d. per lot is reached, to be divided in accordance with the number reared. One brooder was used for each lot, and the cost of oil consumed in lamps was 1·25d. per week each ; that is 5d. for the four weeks. The dry-feeding system had been adopted, and the following report shows the value of that method. The dry food was scattered among the litter, and the birds have to scratch in

finding it, thus obtaining constant and beneficial exercise.   The
following foods were employed during the experiment :—

A.—Dry Food Mixture.   (First four weeks).

	By weight.
Wheat (cracked)...	3 parts.
Dari	2 ,,
Canary Seed	2 ,,
Oatmeal	2 ,,
Millet	2 ,,
Broken Maize	1 part.
Hempseed or Buckwheat	1 ,,
Rice	1 ,,
Meat	1 ,,
Grit	1 ,,

One part of 7 lb. will make 1 cwt.   Cost, 10s. 8d. per cwt. ; 1·14d. per lb.

B —Dry Food Mixture.   (After four weeks).

	By weight.
Wheat (cracked)...	3 parts.
Broken Maize	2 ,,
Dari	2 ,,
Buckwheat	2 ,,
Rice	1 part.
Hempseed	1 ,,
Meat	1 ,
Linseed	1 ,,
Grit and Oyster Shell	1 ,

Cost, 7s. 6d. per cwt. ; 0·8d. per lb.

C.—Soft Food.   (After eight weeks).

	By weight.
Barley Meal	4 parts.
Toppings	,,
Meat	1 part.

Cost, 7s. 9d. per cwt. ; 0·83d. per lb.

D.—Biscuit Meal.

Spratt's parent chick meal, 18s. 4d. per cwt. ; 2d. per lb.

E —Wheat.

7s. per cwt. ; ¾d. (0·75d.) per lb.

The prices charged are those at which the above foods can be purchased generally.

In feeding, the soft food was supplied in sufficient quantities to be
cleared up at once.   The hard corn was left for about half to three-
quarters of an hour, and then removed.

Green food was supplied, but when the birds were out on the grass
it was seldom eaten.   It was given after hard food, so that the weight
of food consumed could be arrived at.

Experiment No. 1.—White Wyandottes.

The average gain in weight in the first four weeks, including the loss of a bird, which died on 15th April, was 5·8 oz.; in the second four weeks, 10 oz.; and in the final five weeks, 15·7 oz.; and that the average cost per bird in the first four weeks was 0·92d.; in the second four weeks, 2·4d.; and in the final five weeks (inclusive of grit for the entire period), 3·5d.

The weight of the thirty birds, when 24 hours old, was 2 lb. 8 oz. At the close of the experiment, the twenty-nine birds, then 13 weeks old, weighed 59 lb. 10 oz.; so that the average gain in weight was 1 lb. 15½ oz. With regard to their respective weights at 13 weeks old, twenty-nine birds averaged 2 lb. 1 oz. The fourteen cockerels averaged 2 lb. 2 oz., and the fifteen pullets averaged 2 lb. The greatest gain was 2 lb. 6 oz., and the least gain 1 lb. 12 oz.

The birds varied in weight on 5th July from 29 oz. to 39 oz., as follows : Five weighed 29 oz. each; one 30 oz.; four 31 oz.; four 32 oz.; four 34 oz.; three 37 oz.; two 38 oz.; and one 39 oz.

While the cockerels made the greatest gain, exceeding the average by 0·74 oz., the pullets were not so far behind as might have been expected, only falling below the average by 0·7 oz.

Experiment No. 2.—Faverolles.

The average gain in the first four weeks was 6·02 oz., in the second four weeks, inclusive of the bird which died on 15th May, 9·8 oz., and in the final five weeks 16·7 oz.; and that the average cost per bird in the first four weeks was 0·95d., in the second four weeks 2·0d., and in the final five weeks (inclusive of grit for the entire period), 3·3d. The weight of the thirty birds, 24 hours old, was 2 lb. 5 oz. At the close of the experiment the remaining twenty-nine birds, 13 weeks old, weighed 61 lb. 11 oz., or an average gain of 2 lb. ¾ oz.

The birds varied in weight on 5th July from 26 oz. to 42 oz., as follows : One weighed 26 oz.; two, 27 oz.; two, 29 oz.; one, 32 oz.; three 33 oz.; seven, 34 oz.; four, 36 oz.; five, 37 oz.; one, 38 oz.; two, 42 oz. each. As to their respective weights at 13 weeks old twenty-nine birds averaged 2 lb. 2 oz. The fifteen cockerels averaged 2 lb. 2 oz., and the fourteen pullets averaged 2 lb. 2 oz. The greatest gain was 2 lb. 9 oz., and the least gain 1 lb. 9 oz.

In this experiment the cockerels did not grow as quickly as the pullets, falling below the average by one-third of an ounce, and the pullets exceeding the average by a little more than one-third of an ounce.

Experiment No. 3.—Buff Orpingtons.

The average gain in weight in first four weeks was 5·6 oz.; in the second four weeks, 10 oz.; and in the final five weeks, inclusive of the bird which died 16th June, 18·0 oz. ; and that the average cost per bird in the first four weeks was 1·4d. ; in the second four weeks, 2d.; and in the final five weeks (inclusive of grit for the entire period), 3·2d. It will be seen that the greatest growth was in the hot and dry week ending 31st May, and the next greatest in the cooler and moister week ending 21st June, while the least average growth after the first week was in the week ending 17th May, when cooler conditions prevailed. The weight of thirty birds, 24 hours old, was 2 lb. 6 oz. At the close of the experiment twenty-nine birds, 13 weeks old, weighed 63 lb. 2 oz., showing an average gain in weight of 2 lb. 1½ oz. The birds varied in weight on 5th July from 24 oz. to 39 oz., as follows : One weighed 24 oz. ; one, 27 oz. ; three, 29 oz. each ; one, 30 oz. ; three, 31 oz. ; three, 32 oz. ; four, 33 oz. ; two, 34 oz. ; two, 35 oz. ; one, 36 oz. ; five, 37 oz. ; one, 38 oz. ; two, 39 oz. each.

In this case it was found that twenty-nine birds averaged 2 lb. 3 oz. The eighteen cockerels averaged 2 lb. 3 oz., and the eleven pullets averaged 1 lb. 13 oz. The greatest gain was 2 lb. 6 oz., and the least gain 1 lb. 7 oz.

In this experiment the cockerels exceeded the average by ·38 oz., whereas the pullets fell below the average by 3·83 oz.

Experiment No. 4.—Cross-bred Fowls.

This lot of thirty birds consisted of fifteen Houdan-Buff Orpingtons, and fifteen Indian Game-Buff Orpingtons.

The cross breeds did not grow nearly so fast as either of the pure breeds, but there was an equal reduction in quantity and cost of food consumed. The average gain in the first four weeks was 4·3 oz. ; in the second four weeks, 10·2 oz. ; and in the final five weeks, 13·07 oz. ; on the average cost per bird the first four weeks was 8d. ; in the second four weeks, 2d. ; and in the final five weeks (inclusive of grit for the entire period), 3d. All the birds in this lot were reared, and thus the average gain should have been greater, but several, two especially, made no growth for part of the time, and thus reduced the average considerably.

The weight of thirty birds, 24 hours old, was 2 lb. 5 oz., which was increased at the close of the experiment to 56 lb., showing an average gain of 1 lb. 12¾ oz.

The birds in this experiment varied greatly from 19 oz. to 37 oz., as follows : One weighed 19 oz. ; one, 21 oz. ; one, 22 oz. ; one, 23 oz. ; one, 24 oz. ; one, 26 oz. ; five, 27 oz. each ; one, 28 oz. ; three, 29 oz. each ; five, 32 oz. each ; two, 33 oz. each ; three, 34 oz. each ; two, 36 oz. each ; three, 37 oz. each ; thus emphasising the importance of selecting the right breed or cross for attainment of early maturity.

The average weight of the thirty cross breeds was 1 lb. 14 oz., thirteen cockerels averaging 1 lb. 15 oz., and seventeen pullets, 1 lb. 13 oz. The fifteen Houdan-Buff Orpingtons averaged 1 lb. 13 oz. ; and the fifteen Indian Game-Buff Orpingtons, 1 lb. 15 oz. The greatest gain was 2 lb. 4 oz., and the least gain 1 lb. 2 oz.

This concludes the particulars relating to each of the lots included in the experiment, and it is now possible to make some comparisons of the results obtained by the different breeds.

Comparisons.

The cost of the chickens at thirteen weeks old is arrived at as follows :—

	Lot 1. 29 White Wyandottes.	Lot 2. 29 Faverolles.	Lot 3. 29 Buff Orpingtons.	Lot 4. 30 Cross Breeds.
	d.	d.	d.	d.
Initial cost of egg	1·48	1·48	1·48	1·43
Cost of working incubator ...	0·21	0·21	0·21	0·2
Cost of working brooder	0·17	0·17	0·17	0·16
Cost of food (average)	6·8	6·75	6·5	5·92
Average cost per bird ...	8·66	8·61	8·36	7·71

In 1904 the cost at twelve weeks old of White Wyandottes was 9·1d., and of the cross breeds 9·09d. It will be seen from the above that this year the cost is less, although the birds were fed a week longer, and that the cross breeds are below the pure breeds, but that is partly explainable by the fact that all Lot 4 were reared. In the above no allowance is made for interest on capital, rent, or labour, as these would vary considerably, and can be calculated by poultry-keepers in accordance with their special conditions. The actual cost, inclusive of eggs, working incubator and brooder, and food, works out as follows :—White Wyandottes and Faverolles, a fraction over 8½d. ; Buff Orpingtons, 8¼d. ; and cross breeds, a little over 7½d.

An interesting point is the comparison of the weights of the birds of each breed at the end of each successive week, as given in the following table. It will be seen that there was very great variation in the growth in different weeks, though, generally speaking, similar fluctuations in growth were made in the same weeks by all the breeds.

Thus the fourth week was a week of considerable growth in each case ; again, the eighth week was marked by a great increase in weight ; while comparatively small increases occurred in the first three weeks, and, again, in the fifth, sixth, and seventh weeks.

From the table we find that while, in pure breeds, both Faverolles and Buff Orpingtons started with a smaller weight than the White Wyandottes, after the ninth week they forged ahead, and stood above at the end of the period. Nearly all the time the cross breeds were in the rear, in spite of the fact that not one of these died.

Comparisons of Weights of Breeds.

TOTAL WEIGHTS.

Ages.	White Wyandottes.	Faverolles.	Buff Orpingtons.	Cross Breeds.
	lb. oz.	lb. oz.	lb. oz.	lb. oz.
24 hours old	2 8	2 5	2 6	2 5
8 days old	3 1	3 3	2 11	3 2
15 ,, ...	4 4	5 8	4 14	4 8
22 ,, ...	6 0	8 4	8 4	6 4
29 ,, ...	13 0	14 0	13 0	10 8
36 ,, ...	16 0	16 11	15 5	13 13
43 ,, ...	19 6	20 2	16 12	17 0
50 ,, ...	23 3	23 13	20 2	20 2
57 ,, ...	31 2	31 6	31 11	31 8
64 ,, ...	37 2	36 6	38 13	35 11
71 ,, ...	41 15	43 3	43 15	41 15
78 ,, ...	48 14	49 0	52 2	44 5
85 ,, ...	51 5	54 7	57 4	49 5
92 ,, ...	59 10	61 11	63 2	56 0

In the next table are given the comparisons of the four lots.

General Comparisons.

	White Wyandottes.	Faverolles.	Buff Orpingtons.	Cross Breeds.
Total food consumed	237·55 lb.	233·85 lb.	226·94 lb.	216·34 lb.
Total cost of food	16s. 6d.	16s. 3¾d.	15s. 8½d.	14s. 9¾d.
Weight of food consumed for each pound gained	4·16 lb.	3·9 lb.	3·72 lb.	4·03 lb.
Average cost of food per bird ...	6·8d.	6·75d.	6·5d.	5·92d.
Cost of increased weight per pound	3·45d.	3·3d.	3·1d.	3·3d.
	lb. oz.	lb. oz.	lb. oz.	lb. oz.
Average gain in weight	1 15½	2 0¾	2 0½	1 12⅔
Average weight (thirteen weeks)..	2 1	2 2	2 3	1 14
Average weight (cockerels) ...	2 1½	2 2	2 3	1 13½
Average weight (pullets)	......	2 2	1 13	1 12½
Greatest gain	2 5½	2 8¾	2 5¾	2 3¾
Least gain	1 11½	1 8¾	1 6¾	1 1¾

As the greater part of the chickens raised were required for breeding stock or for later fattening, they were not killed on 5th July, and thus the gross profit cannot be stated. The experiment, however, shows the actual cost of hatching and rearing to thirteen weeks of 117 birds, as follows:—

	£	s.	d.
29 White Wyandottes, at 8·66d.	1	0	11
29 Faverolles, at 8·61d.	1	0	9¾
29 Buff Orpingtons, at 8·36d.	1	0	2½
30 Cross Breeds, at 7·71d.	0	19	2½
117 Total cost	£4	1	1¾

The total weight of chickens produced at thirteen weeks was 241 lb. 2 oz.

CHAPTER XXXVI.

Faverolles as Layers.

FAVEROLLES having some Asiatic blood in them are sometimes referred to as good winter layers ; however, my own experience of them is that just like other breeds, a good deal depends on the time they were hatched. Early hatched birds will lay early, while those late hatched rarely produce eggs till the following spring. With the infusion of Brahma and Cochin, writers on the breed usually put them down as fowls that go broody regularly. This is entirely a misconception, for in my own yards there is an English, a New Zealand, and a Victorian strain, and of sixteen laying hens last summer only three of them became broody, which, in my case, was a handicap, as broodies of other breeds had to be purchased.

So far as laying is concerned I have kept no records, but fortunately the breed has got a good test at the competition just over at the Hawkesbury College, the only test where they have yet appeared, and here again the date of hatching is an important one in relation to the first year's egg production. The six Faverolles, owned by Mr. Walsk, of Arcadia, were too young at the commencement of the competition in April, 1905, and at the end of two months had not laid an egg, thus commencing with a handicap of one-sixth duration of the test. In the entire 100 pens there were only five other lots with such lee-way to make up, and, one excepted, all finished away far down in the programme. Not so with the Faverolles. Commencing in June with their first eggs, they crept up month by month, ultimately finishing in the thirty-fourth place amongst the hundred, beating a number of pens which had 150 eggs of a start. The Faverolles were laying strong at the finish, and the owner believes, that had the test continued, a further two months, thus enabling the Faverolles to have an actual year's laying period, they would have been amongst the top few. From the time they started they crept up and passed from four to six lots monthly, until on the 31st March, they finished with 1,040 eggs, or 14½ dozen for each bird, weighing 25 oz. to the dozen, and had passed sixty-six pens on the way. Following is the monthly laying of the birds :—April 0, May 0, June 56, July 118, August 140, September 132, October 134, November 98, December 98, January 96, February 89, March 70. Coming to the financial results of the pen, such was also good, although having nothing to show in the two dear months when eggs were from 1s. 6d. to 2s., the total value of the product, was just on 13s. for each hen, beating a number of the more popular and plentiful breeds and varieties. The above was certainly a splendid record, and although in most instances the performance of one pen does not prove much, still in the present case a good deal attaches to the figures, seeing that the competing pens are the direct progeny of the previous year's imported English stock, and prize winning birds at that. The breed is still in few hands here, and has not had time to deteriorate, consequently there cannot be any bad laying strains. My own stock are the result of crossing some

Victorian birds with New Zealand and English, and although not tested, nor yet any records kept, I am satisfied, and the public may be, that those tested fairly represent the laying properties of the Faverolles in Australia. Mr. Walsh, encouraged with the success of his first venture with the breed, is competing with a second lot in the present 1906-7 College competition.

Concerning the laying properties of this breed in other States, Mr. H. May, of New Zealand, from whom I obtained some stock two years ago, wrote me that they were excellent performers in the way of laying, while Mrs. Travers, late of Gippsland, Victoria, from whom my first stock came four years ago, always championed the Faverolles as the best winter layers she ever had. However, private opinion, where no records are kept, is sometimes influenced in favour of the breed one patronises, and even when home tests are made, those in public are more readily accepted, consequently there is no need to go further for the actual performance of the French Dorking than the College records, which show that from a flock of pullets some eight or nine months old, there may reasonably be expected from each hen fourteen dozen eggs in the first laying period.

CHAPTER XXXVII.

Faverolles in Australia.

ALTHOUGH it is only within the past year or two that Faverolles have become prominent in this State, the breed has been known in Australia for half-a-dozen years or more. Perhaps the first arrivals were from a well-known English breeder and judge—Mr. Hawker, who is interested in station property in South Australia. This gentleman on a visit here three years ago, when interviewed at the Royal Agricultural Show, spoke highly of the Faverolles, stating that he had forwarded a number of them to his station property in South Australia, a few years previous.

To Mrs. Travers was due the introduction of the breed to Victoria, her stock being exhibited, and well advertised, secured a number of patrons for them, but, as with most other new breeds, serious defects existed in a number of the stock, much of which has now been overcome. The principal trouble with the early importations was the want of the fifth toe. In the first few years after their introduction to England this was not insisted on, but when the Poultry Club took the matter in hand such was embodied in the standard of perfection, with the result that the Faverolle cock or hen lacking this useless appendage is not eligible for a prize. Mrs. Travers exhibited a number of her birds in Sydney three years ago, and disposed of them to breeders here, but so many of the progeny came with but four toes that people tired of the strain. Prior to the above a medical gentleman of Sydney, now deceased, received some English importations, and although of better colour and larger than the Victorian birds, had not the extra foot embellishment.

Then came some New Zealand birds from the yards of Mr. H. May, and these being the progeny of more recently imported English stock were of the correct colour and more in accordance with standard requirements. These and some later English importations, and a further New Zealand consignment to Mr. H. M. Hamilton constituted the bulk of the breeding stock of this State.

Imported Faverolles Cock.

Sire of the twelve prize winners, Royal Agricultural Society's Show, 1906.

During the past year quite a number of the male birds were sold for crossing purposes, the experience so far being that the bulk of the buyers purchased this breed more for commercial purposes than the show pen.

Reverting to the Victorian stock it may be mentioned that Mr. A. Masseran, of Victoria, visited England and France a few years ago, returning with a quantity of the Salmon and Black Faverolles. This gentleman being experienced in French methods, bred the birds for table purposes, and at the first time of exhibiting in the Victorian shows, won the bulk of the Government prizes offered for table poultry. Mr. Masseran's success in this branch of the poultry business,

Faverolles Cockerel.

First prize, Royal Agricultural Society's Show, 1906. Seven months old; weight, 7½ lb.

secured for him the contract for the supply of the Governor-General's table poultry, Faverolles forming a large portion of the supply.

The writer's experience with this breed has been exceptionally good. Each year sixty or seventy were hatched, and, except through accident, all were reared. It is, however, in the matter of growth that they excel many other varieties. If plenty of food is supplied they grow all the time; and, whether attributable to strain or the general character of the breed, they certainly beat most sorts in putting on flesh and weight.

Faverolles Cockerel.

Five months old; weight, 6 lb.

The illustrations will show the weights at various ages, which are extraordinarily good, but it is in the beautiful quality of the meat that the most merit lies. The appetising appearance of the trussed birds, and the delicate flavour, being all that connoisseurs could desire.

As has been said previously (page 117), there are other colours of Faverolles, namely, Ermines, and Black. Of the latter, Mr. Masseran has been the only importer, one pair of this stock being now in the writer's yards; but of whatever colour, all have the same economic qualities, even the crosses from them possessing good meat quality.

In connection with crossing, the following reply to a correspondent in a late issue of the English *Feathered World* will be of interest :—"Faverolles chickens are excellent birds, and are very hardy. I have seen really tip-top table birds bred from a Faverolles cock and Buff Orpington hens; the chickens were nearly all white,

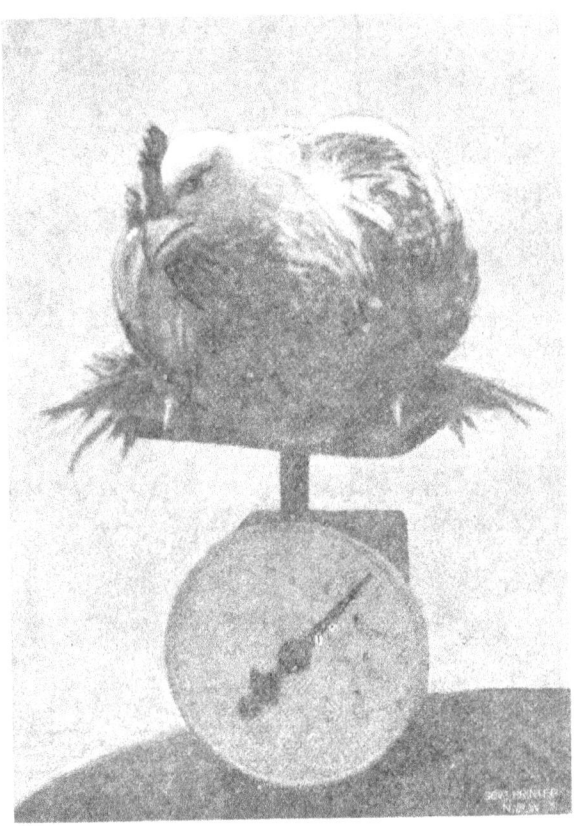

Faverolles Cockerel.

Three months old; weight, 3½ lb.

with just a few striped feathers in the hackle. At four and a half months' old they were a rare size, and the flesh was beautifully white and of good texture. Being a white-plumaged bird, they trussed remarkably well for the market."

It will be remembered that Mr. H. Cadell exhibited several of the above cross at the 1905 Royal Show in the table poultry classes, the colour being, as stated above, almost white, and the carcases were of large size, and covered with a great quantity of white meat.

Enough has now been shown as to the merits of the Faverolles for commercial purposes, while as exhibition birds they should command a good deal of attention from the fact that very little trouble is

Pen of Faverolles Pullets.

necessary with them for the show pen, neither washing, grooming, or trimming being required. The fine show which appeared at this

Faverolles Cockerels at Home.

year's Royal Agricultural Exhibition were just lifted out of their runs and taken to the show.

Appended is the standard for judging Faverolles :—

General characteristics of the cock :—

Head and neck.—Head, broad, flat, and short, free from crest. Beak, stout and short. Comb, upright, single, medium size, four to six neat serrations, free from coarseness or any side work. Ear-lobes, small, hidden by muffing. Wattles, small, fine in texture. Beard and muffing, full, but the beard should be short. Neck, short and thick, especially near the body, into which it should be well let in.

Body.—Body, thick, deep, and cloddy. Breast, broad, keel-bone very deep and coming well forward in front, but not too rounded. Back, flat, square, very broad across the shoulders and saddle, and of fair length, but not so long as in the hen. Sides, deep. Wings, prominent in front, but small and carried closely tucked to body.

Tail.—Carried rather upright, feathers and sickles stout and medium length ; long thin flowing tail feathers, carried low or straight, are very objectionable.

Legs and feet.—Thighs, short, wide apart, plenty of body between them. Shanks, medium length, and stout, straight, sparsely feathered down to outer toe. Knees straight ; carried well apart ; narrowness or tendency to be in-kneed very objectionable.

Toes.—Five in number, the fifth toe clearly divided from the fourth toe, outer toe sparsely feathered.

General shape and carriage.—Active and alert.

Size and weight.—Large cocks, 7 lb. to 8½ lb. ; cockerels, 6½ lb. to 7½ lb.

General characteristics of hen:—

Head and neck.—Head, beak, ear-lobes, wattles, beard, and muffing, as in the cock.

Comb.—Similar to the cock, but much smaller and very neat, and fine in texture.

Neck.—Short and full, carried straighter than in the cock.

Body.—Generally longer and deeper than in the cock.

Breast.—Deep, full and prominent, keel-bone longer than in the cock.

Back.—Broad and flat, longer than in the cock.

Tail.—Fan-shaped, feathers broad, stout, and medium length, carried midway between upright and drooping.

Legs and feet.—As in the cock.

General shape and carriage.—Active and alert.

Size and weight.—Large hens, 6 lb. to 7 lb. ; pullets, 5 lb. to 6½ lb.

Colour in Salmon Faverolles :—

In both sexes :

Beak.—Horn or white.

Eye.—Grey or hazel.

Comb.—Red.

Face, ear, lobes, and wattles.—Red, both partially concealed by muffing.

Shanks and feet.—White.

In the cock :

Beard and muffing.—Black, ticked with white.

Huckles.—Straw.

Back and shoulders.—A mixture of black, white, and brown.

Breast.—Black.

Wing bows.—Straw colour.

Wing bar.—Black.

Secondaries.—Pure white on the outer edge of feathers and black on the inner edge and tips.

Primaries.—Black.

Thighs and underfluff.—Black.

Tail.—Black.

In the hen :

Beard and muffing.—Creamy white.

Head and neck hackle.—Wheaten brown, striped with same colour of darker shade.

Back and shoulders.—Wheaten brown.

Wings.—Similar to back, but the colours are softer and lighter.

Primaries and secondaries.—Wheaten brown.

Breast, thighs, and fluff.—Cream.

Tail.—Wheaten brown.

Value of points in Faverolles—Cock or Hen.

Defects.						Deduct up to.
Bad comb	...	...	...	..	...	10
Insufficient beard or muffing	...	...	...	...	...	20
Defective colour	..	...	...	...	..	25
Want of symmetry	...	...	...	...	..	20
„ size	...	...	...	...	...	15
„ condition	...	...	...	...	...	10
A perfect bird to count			...	...	...	100

Serious defects for which birds should be passed :—

Skin and legs other than white ; absence of all beard or muffing.

Chapter XXXVIII.

MADE IN AUSTRALIA.

As has been shown in previous chapters, a few breeds of fowls, notably Game and Dorkings, have been known in England for generations, their origin and ancestry being difficult to determine, nor does it matter now how they originated. Numerous other breeds and varieties, plentiful and popular, have appeared within the past twenty or thirty years, America and England being responsible for their advent.

These now acknowledged breeds were all the results of crossing from and with the older ones. For a few years they were comparatively nondescript in appearance. However, the patience and preseverance of fanciers brought them to such a state of perfection in colour and type, that they now breed as true as do those whose origin is lost in antiquity.

So far as Australia is concerned, the same material was available as in the Old Country for making breeds, but few troubled with the matter, as years of patient experimenting are required for the completion of such a task, with the result as new breeds appeared in other countries, and became established there, Australians, rather than manufacture, adopted the system of purchasing, and perhaps improving, the ready-mades of other countries.

That this method has been a satisfactory one is evidenced in the show-pens of to-day, the mammoth Orpingtons, artistically coloured, and mathematically-marked Wyandottes, with all their commercial merits, leaving but little more to be desired by the most exacting fancier, farmer, or other breeder. Still the same thing could have been as safely said before the manufacture of the above two breeds. The Langshan was a meat and egg fowl, while others possessed favourable commercial essentials, all of which points to the conclusion that the poultryman's desire for something new is never satisfied ; and should present or future experiments demonstrate that the possibilities of crossing are not yet exhausted, and something newer, if not better than we have, be produced, there is not a doubt that breeders will give it the usual hearty welcome, and if such an origination takes place in Australia, and such be duly and prominently brought before the public, patriotism alone should be a guarantee of a successful future for the Australian fowl.

When writing on this subject in 1897, I then showed that Australian Game were the outcome of crossing the Old English Game and Malays, the object being, not to get an exhibition bird, but, rather, one superior in fighting qualities, hence what is known as Australian Game were more of an evolution than a manufacture. However, something more definite in the matter of fowls made in Australia was mentioned in the same *Gazette* as above. This was a blue fowl, originated in a Melbourne suburb by Mr. J. C. Coupe. They partook a good deal of the Langshan type, and were named Royal Blues. A few came to Sydney, and classes were made for them at one or two shows, but all inquiries on the subject now elicit that the breed, if it ever was one, is now extinct.

Imperials.

A few years ago it became generally talked about in Sydney poultry circles that Mr. W. E. Boutcher, of Canterbury, had originated a breed to which he gave the above name. Classes were made for them at some of the Sydney shows, the originator, who made a good

display of the breed, being the only exhibitor. Some time after they appeared in the Hawkesbury College laying competition, and though the result from one pen does not determine much, still, as this new breed had but a year or two's existence, and in the hands of few people there could be but one strain, the termination of the competition showed that the Imperials behaved splendidly, the six hens producing 146 eggs each, followed by Orpingtons with 137, and Leghorns 136.

Following this came the third yearly competition, wherein 100 pens competed, and here again the Imperials distinguished themselves, the six hens laying 949 eggs, or 158 for each hen, this time being one egg behind the Orpington averages, and ahead of Wyandottes and others. The pen finished in the forty-first place, thus making a better performance than fifty-nine of the competing lots. The eggs were also of good size, weighing 25½ oz. to the dozen. It is, however, at the present 1905-6 laying competition wherein the Imperials have not only justified the existence of an Australian-made article, but also their name as well. The competition commenced on the 1st of April last year, 100 pens of six birds each again competing, and at the close of the eleventh month, one pen of the Imperials have laid the grand total of 1,188 eggs, or 198 for each hen, and are within one egg of second place in the 100 lots. Another pen of Imperials also competed, and again the records are exceedingly high, the total for the eleven months being 1,089, or 180 eggs for each hen. The above are the public tests wherein the birds have appeared, and it is safe to say that no other breed or variety has performed so consistently.

Mr. Boutcher contributes the following on his origination :—

"In dealing with the above breed of poultry, I may say its introduction was not to supply a long-felt want in the shape of a general utility fowl, but rather to show what can be done by crossing judiciously. The Imperial is certainly a farmer's bird, being built to suit the farmer's purpose as an egg-producer and good table variety. The breeds used in its construction are all well-known kinds, viz., Golden Wyandotte, Brown Leghorn, Partridge Cochin, and Black Orpington. With the exception of the Partridge Cochin, the remaining parts of its composition are all looked upon as good layers, the last-mentioned being used only to secure the desired colour, which turned out very well. As a layer, the Imperial can stand side by side with the most popular egg-machines we possess, the laying competitions having verified that, whilst the table qualities of the breed rival our best.

"The Imperial was brought before the public in 1901 ; classes were provided for them at the New South Wales P. P. C. and D. Society in 1902, which was copied by most of the leading poultry shows. So far the Imperial has not received the patronage it deserves from the general public.

"In commenting on the new breed, the *Sydney Morning Herald* of 6th June, 1903, says :—' The cock is a handsome and symmetrical bird, weighing about 10 lb. Its rich plumage recalls the bright colouring of the Partridge Cochin, but its dark legs and feet are free from feathers. Its headpiece and general shape suggest the Orpington. The ground colour of the hen is dark, with buff pencilling, the average weight being 7 lb. Classes for Imperials were provided at recent shows. As a table bird the weight of the Imperials speaks for itself. As a prolific egg-producer no better evidence of the value of the breed can be given than the result of the laying competition recently concluded at the Hawkesbury College. This event was a twelve months' test, open to all. Thirty-eight pens, each containing six hens, represented various breeds. When the averages were extended at the termination of the competition, it was found that six Imperials had averaged 146 eggs each. Seventy-six Orpingtons were second with 137 each, and forty-two Leghorns third with 136.'

"In dealing with the same subject, the *Daily Telegraph* of April, 1902, says :—'What may be termed the first successful effort in Australia to originate a breed is that of Mr. W. E. Boutcher, of Canterbury. The breed, to which he has given the name of Imperial, seems destined to yet make a name. Briefly, it is a partridge fowl of the Orpington type, and as a general utility bird it ranks with Mr. Cook's creation.'"

As a general utility fowl suited for farmers or other breeders whose object is profit, the Imperials should fill every requirement. The Leghorn, Wyandotte, and Orpington are a combination of utility, the Cochin blood being responsible for the colour.

The breed has now made a name for itself at the three competitions, the most notable thing being that while many good performers in the early tests have done badly in later competitions, the Imperials have done better each time they tried, all of which goes to show that fowls can be not only made in Australia, but superior ones at that, and when those about to begin poultry-keeping, or others desirous of changing bad performers to good ones, the pocket will be consulted and patriotism gratified by adopting the Australian-made, aptly named Imperials.

———

This now completes the articles on "Farmers' Fowls." A few miscellaneous sorts have not been included, but those desiring the best sorts will find one or all of those dealt with admirably suited for whatever purpose or requirement.

The next chapters will deal with other subjects in connection with the profitable care and management of poultry.

CHAPTER XXXIX.

Location.

HAVING dealt exhaustively with all the breeds considered suitable for the farm, attention will now be given to where and how fowls should be kept, and without further introduction, I may say that the agricultural farm of all places is the ideal one for profitable poultry-breeding. Farms devoted to the sole purpose of breeding fowls for their carcase and eggs, and having to pay rent, taxes, wages, feed, and a number of other expenses, are very rarely a success, and such establishments I warn all to eschew, the many failures of such being largely responsible for the too frequent detracting phrase applied to the industry, namely, " It won't pay." That there are farms devoted solely to the production of eggs and meat, and from which a living, more or less satisfactory, is being made, I freely admit; but each of these has some special advantage in procuring much of the fowls' food at other than market rates. Hotels, restaurants, and boarding-houses are daily visited by some of these breeders, loads of excellent fowls' food being obtained from such places for a trifle beyond that of the carting of it away. Glebe Island also supplies tons of stuff weekly, which, when prepared and cooked, has properties excellent for egg-production. The wharfs and flour-mills are also exploited for occasional cheap lines, all assisting to make this class of poultry-farming a paying one. However, with the above advantages, which are only available to those within easy distance of the city, there are many detracting features. On farms where no other stock are kept, poultry have to be bred in immense numbers to return a profit. The business of cooking, cleaning, feeding, and otherwise attending to these large numbers is not of the pleasantest, for, no matter how clean the place may be kept, or how sanitary the arrangements, disease will in time make its appearance, and to dose and doctor a number of roupy fowls with a prospect of curing is a disagreeable experience. Then there are all the disappointments in hatching, the mysterious dying of large numbers of well-cared-for chickens, and at times unsatisfactory markets, are but a few of the handicaps.

There are no eight-hour days on a poultry farm; from daylight to dark in summer, while in the winter many hours have to be spent after sunset over the thousand and one things connected with a large poultry plant, and of all other businesses this one is that of seven days in the week; indeed, those whose lives are cast in such places are all eloquent on the hardships inseparable from the calling.

Another variety of poultry-farming, and more satisfactory, with less unpleasantness, is that of the poultry fancier. This comes about by the purchasing of one or more pens of high-class pure-bred fowls of some popular breed, if good enough to exhibit and occasionally win. There is, with judicious advertising, a considerable sale of eggs for hatching, and in season a demand for pure-bred birds at prices considerably beyond that of the ordinary market sort. This branch of the poultry business is carried on to a considerable extent in the suburbs of Sydney, those who go in for this branch being usually people who are in business in the city, and to many at the present time it affords a considerable adjunct to their ordinary wage or salary, while in more than one known instance the sale of stock and eggs so increased that the owner gave up his legitimate calling, and confines himself to the breeding of fancy fowls only. Several at the present time are doing fairly well at this branch; but, taken as a whole, the fancy requires some side issue to assist it.

That other branch of the industry—poultry on the farm—is the chief object of these articles, and whatever the handicaps to profitable breeding, as shown above, the keeping of fowls on a farm is subject to none, and is the most pleasant part of the business to write of, for no matter what way the subject is looked at, we must come to the conclusion that the agriculturist, of all others, is the man who can and should make poultry-breeding a profitable undertaking. He has the land, for which he pays no rent for the fowls, the looking after them is within his own family, while for food, no matter what crops may be grown, there is always waste—unmarketable cereals, roots, &c.—which actually cost nothing, and can be profitably fed to the fowls. Then the fowls on a farm usually have a free range, and gather up large quantities of natural food, such as weeds, seeds, and numberless insects, all of which contribute to healthy and cheaply-fed fowls. Poultry keeping on a farm can be carried on at a minimum cost, and this despite the fact that the location may be far removed from the Sydney markets.

In an earlier portion of these articles, I showed that we got our birds from England, and judged them at the shows in accordance with the English standard, a feature to which none care to take exception. We, however, do more than that; we get the bulk of our poultry books and poultry literature from the same source, and, unfortunately for ourselves, slavishly follow the advice given therein as to management, forgetful of the fact that the climatic and other conditions are so different that what might be the correct thing for the cold, damp climate of the United Kingdom would be altogether unsuitable for this semi-tropical country.

In connection with location, we are told to select a high, dry place for the fowl-runs, that they must be well drained, and face a certain point of the compass, while for houses, the sort advocated and approved of for English conditions are, in many instances, copied here and with disastrous results; an illustration will suffice. Not long since, I visited what was termed an up-to-date poultry farm, devoted solely to the fancy side of the industry. The holding was, perhaps, 20 acres, and embraced a high, dry, cleared paddock, minus even a solitary tree, the soil, or rather earth, being of a hard clayey nature. This extended down to a very low ground with trees and much undergrowth, with a magnificent soft surface, composed of the organic remains and other débris washed from the higher ground by the centuries of storms. The reputed model poultry plant was built according to English ideas, on the high, dry, hard hill, the runs extensive enough, but not a leaf of shade, nor a blade of grass; while the expensive and well-built, but ill-ventilated, houses had a temperature on the day of my visit equal to that of an English hot-house or a gardener's forcing frame,—conditions which were certainly responsible for the peculiar noise then being made by many of the fowls, true symptoms of a cold, and forerunner of the frequently fatal roup. These prize fowls when fed in the morning, although having plenty of space, had no occupation or inducement to wander on this selected high and dry place, and just squatted down throughout most of the day waiting for their next meal, and following that, another dreaded night in stuffy sleeping.quarters. This was the week in, week out life of these prize fowls in these prize poultry yards, on which one-tenth of the money spent would have given better results. The one and great mistake was the too faithful adherence to the conditions obtaining in other countries, and particularly to the high, dry, and well drained bogey. I have to add that the farm, as a whole, was an ideal one for the purpose of the owner, *i.e.*, keeping in health, breeding and rearing prize fowls, which would have obtained had the owner erected the houses half-way down the hill, spent less money on their architecture, and extended the runs right down into the lowest portion of the ground, enclosing the low land with its deep, black, soft soil, the trees, bush, and other growth, leaves, &c., these harbour myriads of insects and other of Nature's foods, and affording that great essential to fowls' health—scratching exercise—and thus prompting a better egg production. Shade, also, would be afforded, an important element in this country of perpetual sunshine. Had this portion of the farm been given to the fowls, their life would have been a more natural one, for, despite the fact of supplied foods, centuries of domestication has not yet deprived fowls of their instinct of scratching for a living, this contributing to their contentment and beneficial to their welfare in many ways. When the above was suggested to the proprietor, he replied that Lewis Wright said so-and-so, forgetful of the fact that that authority was never in Australia. It was evidently not apparent to the designer of the plant that there was nothing to drain, and even in the lower portion referred to, a couple of inches of rain in twenty-four hours, although beneficial, would not be visible; indeed, were there .a 6-inch fall in one week on any of our suburban poultry farms, such, although inconvenient at the time, would have other than ill-effect. In relation to the high, dry, and well drained theory, such is certainly applicable to England, with its 250 or more wet, sunless days in the year, and where frequently the roadway is not dry for three autumn and sometimes the three spring months of the year. The soil, whether of poultry or other farms, is mostly wet and cold; while for houses in a country where wind, rain, sleet, snow, hail, and frost obtain so many days in the year, and the sun shines on so few, such cannot be too comfortably constructed. The poultry-farmer and poultry-keeper in Australia is more highly favoured by natural conditions than in almost any other country in the world, the simplest possible structure, to keep off the occasional tropical showers, and a breakwind of some sort, being the only requisites for housing in a large portion of the State; and there are numerous instances of profitable poultry keeping where the birds are allowed to roost on trees or fences, and sometimes prize-winners at that. In a case of my own, at Randwick, three years a hen hatched eleven chickens in a corner of the garden, no overhead covering whatever, the only protection being a paling fence on two sides and some garden growth in front. During the hatching

period over an inch of rain fell in fourteen days; the eggs were coated
with the wet soil after rain, still eleven were hatched out of the thirteen.
The hen had full liberty with the chickens, brooding them wherever she
liked, but never under any constructed covering, and at nine or ten weeks
took them into a lemon tree to roost, and, when at about 3 months old,
over 1½ inches of rain fell in one night, without apparent ill-effect.
During the period, one chicken disappeared; the remaining ten were
reared by the hen, one of these being exhibited at the following Royal
Agricultural Show at Moore Park, in the same class as his imported sire,
the latter highly-pampered oversea aristocrat having to take second place
to the Australian native, bred and reared under open-air conditions, his
only covering being the Australian sky. On two subsequent occasions
the native-bred distinguished himself in the same way, and several of
the other sex in the same brood grew to be bigger and better exhibition
specimens than their English-bred and far-travelled matron.

It must be here distinctly stated that I have no intention of advocating
the above open-air sort of poultry keeping, the purpose being rather to
show that such results would be utterly impossible in England or America,
and that when the success noted was achieved here, still better results
must be expected where dry overhead conditions obtain. The above
experience was in respect to the rearing only; but that the very best
results can also be obtained in egg production by simple methods in
housing and feeding is also overwhelmingly proved, as will be seen in
the *Gazette*, June, 1906, where, at the Rockdale egg-laying competition,
where everything as regards housing, appliances, and feeding was simplicity and cheapness itself, 300 Australian-bred hens, of different breeds
and varieties, made a world's record by laying the grand total of 58,736
eggs in the twelve months, or slightly over 195 for each hen.

Chapter XL.

Houses, &c.

Coming to the structures best suited for Farmers' Fowls, the chief consideration will, necessarily, be cheapness, consistent with practicability;
and it is on the farm and about the farm-house where the above requirements are most easily had.

I should first say that the best results will be obtained by open-fronted
buildings, the ordinary lean-to house—a back, two ends, and open front,
and overhead covering—being all that is necessary. The most inexpen-
sive protection can be erected in some corner of the farm yard, the
ordinary paling fence forming the back and end of house, leaving but
one end to erect, the front being open; the further this end is kept from
the right-angle fence the house will, of course, be the larger, and accommodation can be had for any number of fowls in relation to the distance
this end is kept from the fence which forms the other end. The roofing
timber, such as joists, &c., can all be had on the farm, while for the roof,

cheapness being the object, bark can be had on most farms, and while admitting that such may not make a picturesque fowl-house covering, still, for practicability and comfort, it suits excellently; nor is it anything to be ashamed of, seeing that this product of the bush has formed the parental roof of many present-day happy and prosperous farmers. At the same time, although the bark roofing be watertight, this and the paling walls will not constitute a healthy hen house, arising from the fact that such a structure would be most disastrous to the inmates by the draughts coming through the openings between each paling; indeed, rather than recommend this rain-proof structure, it could be safely asserted that the fowls would be more healthy roosting on the fences in the open air. The most dreaded of all poultry diseases is roup, of which a cold is the forerunner, the latter in many instances being induced not by the lowness of the temperature, for while a single specimen, or a

A cheap fowl house, wire netted front, bark roof.

whole flock of poultry, may be healthily and profitably kept in open-air conditions below freezing point, this same flock if in a draughty roosting-house colds would be induced at any temperature from 50, 60, or more degrees; and, on the other hand, the house may be so stuffy, close, and otherwise insanitary, that the almost tropical temperature of 80 or 90 degrees F. has been responsible for roupy troubles. To make the above improvised fowl-house draught proof, such can be readily done. On almost every homestead there are a number of old waste corn bags or sacks. These can be cut open, and will make an excellent lining for the house, and should be tacked right round, and from the floor up to considerably above the perches, which should not be more than at most 2 feet from the floor. It may be rightly said that this lining would be a harbour for vermin; however, that is easily provided against. The following recommendations, if carried out, will be effective in making the place vermin proof:—To give the inside of the house, the palings, a good coating of coal tar, and tack the bags on before this dries, and another coat of tar on them after being fixed. The latter repeated each season would contribute to a clean and healthy house; but, unfortunately, the tar is not readily obtainable at the bulk of the farms. However, other more readily obtained applications are effective, of which the following, for simplicity and cheapness, cannot be improved on. Everyone knows how to make whitewash, and that it is best made with fresh lime and boiling water. To a bucket of limewash, when hot, add say, a quart of kerosene, mix thoroughly, and give the palings a good coating, the following day a second application; after this the bags can be fixed, finishing up by giving these a good coating of whitewash, which will

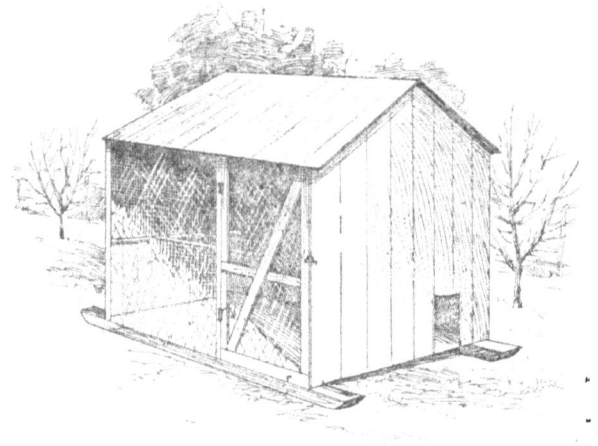

A movable fowl house on runners.

contribute to clean, healthy, sanitary quarters. The perches, also the product of the farm, should get an occasional painting with kerosene, and the inside of the house a whitewashing at least once a year. In districts where native cats, foxes, or other enemies of the poultry yard prevail, the entire front of house, including door, should be wire netted, thus providing that the fowls could be secure at night and kept safe from the, at times, serious depredations of the above. A roosting-house constructed as described, the materials of which are obtainable on the farm, the lime and kerosene excepted, costs almost nothing for labour, materials free, will last for years, and, if other conditions be right, will house and accommodate a flock of prize or other fowls more comfortably and healthily than is done in the buildings constructed under the supervision of the most competent architect, whose fee alone, on what is known as up-to-date poultry plants, would be more than the value of the entire stock of many a farmer's fowls.

However, there are many of those on the land who make a point of not only having their dwelling architecturally and artistically correct, but the farm offices or out-houses as well, and to those whose taste and purse tends to that end the several drawings of poultry houses which appear in English poultry works and journals will appeal, and from which a selection can be made, always remembering that the bulk of them have been designed for colder climates than ours, and to adapt them to Australian conditions will involve considerably more ventilation than is provided in the majority of them.

The various designs are of tongued and grooved boards, would require a carpenter to make, and would cost from 30s. up to as high as £5. In the United Kingdom, one of the most profitable branches of poultry breeding is that adopted by farmers in hatching a quantity of

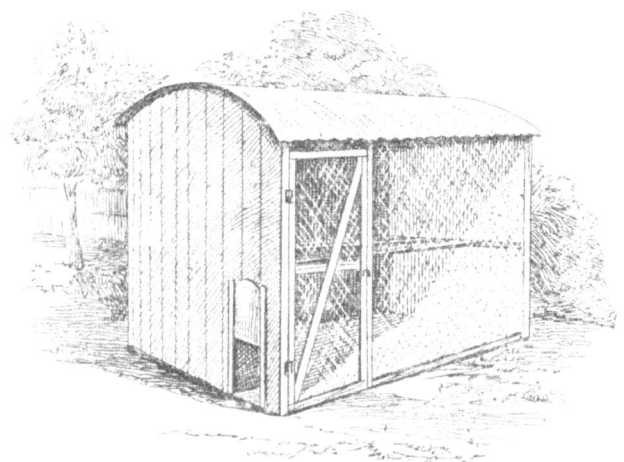

A serviceable fowl house, curved iron roof.

chickens during the latter end of May, June, and July. This, although late to realise the best price at selling time, is more profitable than the early ones withal. These chickens are fairly well grown at harvest time, and, when the wheat and oats are removed to the farm yards from the fields, the hens and chickens are turned out to the stubble, where they find almost their entire food from the grain which has been shaken out of the ear by the wind, or during the operation of harvesting. Of later years, manufacturers have come to the farmers' aid in this respect, by putting on the market what is known as portable houses, some on wheels, and others on slides. These are, in reality, a camp-out for the fowls. These houses are filled to accommodation at night with the half or three-quarter grown chickens, and, in some instances, laying hens. A horse is yoked to it in the morning, and the establishment removed to some portion of the stubble field; the house is opened in the morning, when the inmates betake themselves to the stubble for their grain and insect breakfast, and continue throughout the day looking for and finding a living, the fowls, even the first night, going readily to their new quarters. Should 'here be much cast grain on the ground, the house is allowed to remain several days at this spot, while in the event of it being a poor feeding ground the house is removed each morning to a more rich feeding place. In some instances, supplied food is given once a day, in others, the fowls thrive and put on flesh on what they gather. In this country, such a profitable practice does not obtain; possibly our wheat growers are in such a big and prosperous way that they think fowls are too small a thing to trouble about. At the same time, realising that this State paid considerably over £30,000 last year for imported eggs, a portion of this would surely not be objected to when such could be assured by the small efforts of building a few portable houses and purchasing a stock of young laying hens. But, even did the wheat growers object to be bothered about hens, there are always the farm labourers and other wayside residents who, with the initial low-cost portable houses, and stocked with even the too common sort of fowls, could readily get permission of any of these large stubble areas, the laying, growing, and fattening from this free-food system being largely all profit. Indeed, with the extraordinary acreage in some of these New South Wales wheat paddocks, a large poultry farm on wheels could obtain for almost three months in the year. A variety of poultry houses as used in England are shown, but with a wire-netted front to suit the altered conditions of this country.

After houses come the coops, setting-boxes, &c., the latter, on the duly recognised prize poultry farms, being quite a feature. The farmer, however, requires very few or none of these. The farm-yards and out-buildings afford so many secluded corners that quite a number of broody hens can be suitably accommodated for their three weeks' retirement without one penny outlay. However, when the chickens arrive, the question of accommodation arises, and whether a free range or confinement for the hen and brood is best, there is no disputing the fact that the first three or four weeks are the most important in a chicken's life; and despite the fact that the free range in the farm-yards, with its multifarious weeds, seeds, and insects, affords an ideal chicken existence, the advantages of such will be of more consequence after the above age. There are still some who consider that, where circumstances allow, the hen and chickens should have entire liberty from the day of hatching. This system I can absolutely condemn. There are certainly some hens that are most careful, and know that for a time their chicks require much brooding, heat, and rest, but others, if given their liberty, run the chickens off their legs, frequently resulting in many losses; indeed, the bulk of practical poultry-breeders are now satisfied that to rear a brood successfully it is necessary that the hen be cooped up for three or four weeks, the chickens of course to have liberty. They will, however, up to this age not wander too far, while for the first week or two the sense of confinement, and having little scope for exercise, encourages the hen to brood the chickens more frequently, and thus impart heat to them—that great essential to young life. Absolute confinement is, however, neither advocated nor yet necessary; indeed, after the first three or four days the hen and chickens should be allowed, say, half-an-hour or more liberty for dusting, which is actually the fowl's bath, instinct thus prompting them to this means of ridding themselves of that bane of all living animals—vermin. I have mentioned a half-hour's exercise; this should not be extended, otherwise, when re-cooped, the hen becomes fretful, pines for liberty, and will not brood the chickens in her former way.

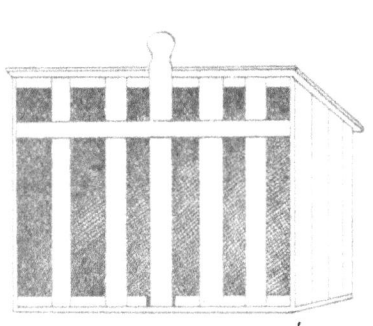

Fig. 1.

The above are, perhaps, reasons enough to show that, for success in rearing, it will be best to coop the hen. The next consideration being the sort of coop, here again the advantages of the farm are apparent. In or about almost every holding there are old cases, boxes, and other grocers' empties, which the farmer can, with a hammer and a few nails, make adaptable to the most extensive requirements. No wooden floor is

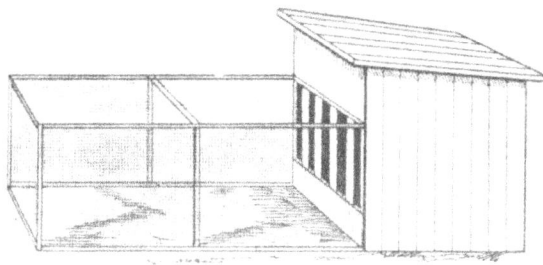

required, the earth being best, the front of box taken out and a few laths substituted, with a space of from 2 to 3 inches between each to allow the chickens to come out and in. The coop should, in the interest of cleanliness, be moved a short distance each day, and which also warrants the chickens venturing on to fresh ground in their wanderings. After four or five weeks' cooping the hen and chickens can be given full liberty.

Just as with poultry houses, some farmers may prefer appearance, and are prepared to spend a few shillings on more modern constructions than those mentioned. To those who do, there are quite a large number of designs to select from; but, personally, I have found that illustrated Fig. 1 is the most easily and cheaply built, neat and adaptable in every way to the requirements. The coop is 2 feet square, made of $\frac{1}{2}$-inch tongued and grooved lining boards, 8 feet of inch square stuff for cornering, and a few laths for front. The wood for the complete coop will cost under 2s., and can be built in a short time. Two of this construction I have had in use for the past eight years are still sound. Some of the other coops illustrated will commend themselves for inexpensiveness, while those who desire neatness in their poultry outfit can make a selection from the illustrations submitted.

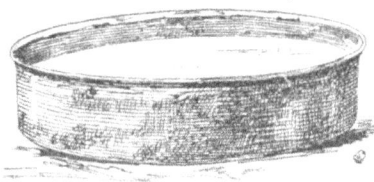

Herring tin.   Drinking Vessels for Chickens.   Small meat tin.

Following coops come drinking vessels. The latter in poultry supply houses are quite a feature, made of either iron or crockery, and of many designs. For the poultry fancier or other breeder whose taste is for neatness, and has a lengthy purse, there are a multitude to select from. The ordinary farmer, however, requires none of these. The usual kitchen empties afford designs adaptable to every stage of poultry-rearing, from the newly-hatched chicken to the adult fowl. There are a small number of poultry-breeders

Large meat tin.   Sardine tin.
Drinking Vessels for Chickens.

who adopt a system of withholding drinking water from the chickens for a period of from one week to several. The fact, however, remains that immediately after chickens have their first feed, whether it be dry meals, broken grain, or moistened soft food, if water is available they drink, and it must be admitted that at this stage of life drinking is neither an acquired habit or taste, but a prompting of instinct, and realising this the wise poultry breeder will place drinking water before the brood at all times, and in suitable vessels; by suitable I mean those easy of access, and not too deep, otherwise the chicks may get into it and receive a chill, if nothing more serious. To meet chickenhood requirements in this respect, I have found nothing equal to the ordinary sardine tin, what is known as halves being the most suitable. These are quite flat, which prevents them being capsized, and should the chickens trample into them, the water is so shallow as to have no ill effects. When the brood gets two or three weeks old a larger bulk of water will be necessary, the round or oblong herring tin becoming the more suitable; later on larger vessels will be required, and such can be had in abundance about any farm-house or other dwelling. The 2-lb. meat tin, the fruit-can cut down, and a host of others are all equally adaptable, and the cost nil. For a water vessel for adult fowls the kerosene tin is *par excellence*. This article, as will be seen by the illustrations, can be cut into a variety of forms, and all as suitable as the most expensive water fountains, whether of American or other designs.

The Kerosene Tin in the Poultry Yard.

The kerosene tin, unlike the others, is not altogether a waste article, still its plentifulness is such that even when it has to be purchased the price is from 2d. to 3d. each, being about one-quarter its original value, and there is not a doubt were it three or four times the present price it would still continue the most universally used vessel on the poultry farm—for in addition to the above, with the ordinary wire or old bucket handle it is used for conveying water to the fowl-yards, for heating water for mixing the morning foods, for cooking the waste vegetables, meat scraps, &c., for carrying the food to the various runs or yards, and as a vessel used in cleaning out the poultry houses. With one side taken off it can be used for mixing the food for from thirty to forty fowls. Taken altogether, as has been shown, poultry houses and coops on the farm cost but a trifle, while the other requirements in the way of appliances are as a rule to be found in the scrap heap. This further shows the advantages the farmer has for keeping fowls over those who have first to debit the undertaking with a large outlay for houses, runs, appliances, labour, and rent, and if the latter make the business pay, how much more so should the farmer whose labour is in his own family, who has no rent to pay for this side issue, and in the matter of food, not a day in the year passes but something is available from the land, for which the ordinary poultry-farmer has to pay, all of which emphasises the contention that with good available metropolitan markets for live poultry and eggs the one class of the community who should do well is the farmer, or other landholder of the State; and realising the many facilities and advantages in their favour, should these people not in the near future very materially increase their now limited output of this article, there is just one, and, perhaps, a wholesome conclusion—they are too well off to bother about fowls.

Chapter XLI.

Feeding, &c.

Of the numerous issues in connection with profitable poultry-keeping, none of late years has received so much attention as aliment. This cannot be wondered at, seeing that the difference between the cost of what a fowl eats and what its carcase or eggs realise is the chief end of all poultry-keeping, and termed profit; and how to reduce the cost of keeping a hen, and thus increase her profitableness, continues to afford a big field for scientific investigation. In the forefront of these are the numerous American experiments, where most exhaustive tests have been made with every known food available in that country, and despite the fact that the conclusions are not all unanimous, there is the satisfactory declaration from all that the ordinary cereal foods obtainable or grown on a farm, and which have been used for ages as fowls' food, cannot be superseded either for cheapness or profitable results. It is to these American stations that we of late have heard so much of the wide, the narrow, and the balanced ration, the proportion of the nutritive elements in the various grains or other foodstuffs, and the lengthy scientific articles on the subject which appear in the American poultry journals, and frequently copied here, and which appeal to the medical fraternity, the analyst, the scientists, and teachers and students of chemistry, and others who understand them, the articles are valuable in the extreme; but to the ordinary Australian poultry man or fancier whose first lesson in chemistry is long since forgotten, the bulk of the science of feeding articles are so much waste.

There is no desire here to discount in the remotest degree the value of the investigator, but when in a one-inch space of an ostensible utility poultry journal, we find the following, it will be conceded that such terms are neither appreciated nor applicable to Australian poultry-farms:—" bile," " calories," " carbon dioxid," " maltose," " zylose," " scrum albumin," " petone," " asparagin," " gliadin," " pentosans," &c., and numberless other as obscure terms are given as necessary knowledge in the feeding of fowls.

The fact, however, remains, that the strenuous life of the poultry-farmer in his efforts to make a living from his fowls has neither time nor desire to master all that is involved in the above terms, particularly as he has had overwhelming evidence that the best results ever obtained in any country from fowls have been in this State by the simple bran, pollard, and wheat judiciously supplied. Less than three months ago I had a call from a gentleman who could discourse on the chemical elements of nutrition for poultry in the most scientific language, yet he wanted to know how it was his birds did not lay. On inquiry I learnt from him that the fowls were at his place when he purchased it three and a half years prior to his visit, and he did not know what age they were at that time, and was unaware that feed was of the smallest consideration so far as his stock was concerned, the bulk of the birds having laid all the eggs they ever would lay, irrespective of scientific compounds.

The fact obtaining that a number of young hens of any breed of, say, from 6 to 9 months old, if turned into the primitive bush will find weeds, seeds, and insects, which will not only sustain life but will also be sufficient to furnish a moderate egg supply, while the same number of aged hens if kept under the most sanitary conditions in regard to housing, and fed on the approved balance ration for laying, will produce but very few, for the simple reason that their ovaries are exhausted, and they have no more eggs to lay. However, irrespective of what is said above, there are a few simple scientific facts in connection with food and feeding from which the poultry or other farmer can derive much benefit, and which is given herewith in the simplest form possible. At the same time, whether the salient points of such be committed to memory and the feeding of a flock of fowls be formulated from it, the fact obtains and will remain that in order to get the best results in the way of a generous egg supply, the first consideration will be young healthy fowls of almost any of the now popular breeds. The next and chief of all essentials being that they should be of a good laying strain or family, this having been conclusively proved at every laying test held throughout Australia, the last Hawkesbury and Rockdale competitions being noteworthy instances. At the last College test one pen of six white Leghorns laid 1,411 eggs, or just on 234 for each hen; while at the same test another pen of this breed, housed, fed, and managed exactly the same, produced but 635 eggs, considerably less than half the above. One pen of Silver Wyandottes laid 1,303 in the twelve months, and another pen but 820 eggs. At the Rockdale competition a pen of Black Orpingtons finished at top with 1,461 eggs, being over 243 for each hen; another pen finished in the lowest place with but 928. Scores of like instances could be quoted, all showing that neither breed nor yet feed was responsible for the extraordinary difference of production, family prolificness or strain being the important feature, consequently, although a table of food values are given, the poultry man who mixed his foods on the most perfectly balanced ration, whether for egg production or carcase, might find this a more expensive system than the usual morning meal of bran and pollard, and the cereal most available for the evening food.

We all know that the object of feeding is to provide material to sustain heat, growth, development, energy, and to supply the essentials for the formation of eggs. When more food is given and consumed than meets the above requirements, the surplus goes to fat, and is stored as a reserve, and if conditions be such that this storage does not take place, then the excess food is lost; and to prevent this loss, and a proper utilisation of the food, a few chemical facts may be stated.

The following table gives the classified component parts of foods, and shows how each is made up, and the functions, together with the percentage of each for a properly balanced ration, considered sufficient for health without reserves:—

Appellation.	Constituents.	Functions.	Proportion in every 100 parts for a normal ration.
1. Albuminoids or proteids.	Nitrogen, oxygen, hydrogen, and carbon.	Flesh, blood, tissue, bone, and egg formers.	13·5
2. Fats or oils	Carbon	Fat formers (both for body and egg) and heat producers.	5·2
3. Carbo-hydrates	Carbon, hydrogen, and oxygen (in the form of starch, sugar, and gum).	Heat producers	55·8
4. Ash	Salts and minerals, such as phosphorus, sulphur, lime, and alkaline salts.	Phosphorus and lime for bone and egg-shells; sulphur for feathers; salts for blood and digestive processes.	3·
5. Husk or fibre	....	Of little value	9·
6. Water	....	....	13·5
			100

The albuminoids or proteins are a group of elements of which the white of an egg or lean meat will suffice for an example. Flesh, blood, feathers—in fact, the bulk of the fowl or animal is composed of this substance. Protein, while building up the animal, is also a heat producer when used in excess. The chief functions of fats are as heat-givers. Carbo-hydrates, milk, sugar, starch, are forms of this element, and produce heat and mechanical force;

when there is an excess supplied it is stored as fat. Ash is the part of food left after burning. The supply of ash in most foods is sufficient for the needs of animals.

Nutritive ratios is the proportion of albuminoids or proteins to the carbo-hydrates, and the proper balancing of these nutrients is of much importance to all stock-keepers. In working out a dietary it is not necessary to take into account ash, husks, or water, and it simplifies matters to treat the fats as carbo-hydrates. What are called feeding standards is the mixing of certain foods in the proportions necessary to produce particular needs, and in poultry there are three standards of requirements, namely, for growing stock, for fattening, and for producing eggs. There are four groups of nutritive substances essential to sustain life, protein, carbo-hydrates, fat, and minerals, and it is the arranging of these in proper proportions that constitute a feeding standard; and these standards have to be prepared for the different classes of work required of them ; in other words, the food has to be arranged for the fowl's requirements, i.e., egg production, growth, or fat. Further than this, the value of any food is in relation to its nutritive digestibility. In vegetables the nitrogenous elements are not present in so concentrated a form as in animal foods, and the non-nutritive elements of all foods are expelled. For the production of eggs and the laying on of flesh the same class of constitutents, the albuminoids or protein, are required. In fattening it forms muscle and lean meat, and in the laying fowls it produces albumen or the white of eggs.

In egg-production protein is a very important factor, seeing that an egg contains 13 or 14 per cent. of protein, hence it will be necessary to provide this feeding element when eggs are the object, the best balanced ration for this object being one part of protein to four or five parts of the other constitutents. This protein is to be had in the most available and cheapest form in lean beef. In a natural state the worms, grasshoppers, and other insects provide this protein. All the cereal foods supplied to fowls contain the same constitutents, the difference between them being that some have a greater proportion of flesh-forming materials—protein—than others. Some excel in heat-producing properties—carbo-hydrates—maize and buckwheat being examples. Should protein be in great excess, fowls suffer in health ; while if the carbo-hydrates are fed too much, fat is produced, and egg production suffers. This brings me to the analysis of grains and other foods, but unfortunately we have here to fall back upon the analysis made in England or other countries, where the climatic conditions are so different from here as to very largely affect the accuracy of these for the conditions here. Take that most popular of all cereals, wheat. The English analyses of a large number of wheats show it to contain 14 per cent. of water, while in tests made on samples grown in our dry climate, the moisture element was but 10, and it is quite conceivable that a like variation would occur in all our cereals.

As an assistance to formulate a diet for the various requirements of fowls, it may be stated that for growing stock a ratio of 1 : 3·54 is considered best, that is, one part of protein to three and a half of the other elements. For fattening stock 1 : 5·6, and for laying stock 1 : 4·55 are given as the correct ratio. Lewis Wright in his great poultry work gives a table of food values, and is often quoted. This, however, is not now much used as the basis of a dietary, the following, the result of more exhaustive experiments, being now more generally adopted:—

TABLE showing the constituent parts in every 100 of the principal Foods suitable for Poultry, with their nutritive ratio.

Foods.	Albuminoids or Flesh-formers.	Fats or Oils × 2¼ = Value in Carbo-hydrates.		Carbohydrates or Heat givers.	Ash (Salts and Minerals).	Husk or Fibre.	Water.	Ratio.
Approximate propor-tion of a normal ration	13·5	5·2	=11·7	55·8	3·	9·	13·5	1 : 5
Oats	13·	6·	=13·5	54·	3·	10·	14·	1 : 5½
Wheat	12·	2·	= 4·5	69·3	1·7	2·5	12·5	1 : 6¼
Barley	10·	2·	= 4·5	58·	2·	14·	14·	1 : 6¼
Maize	10·	5·5	=12·4	66·	1·5	5·	12·	1 : 7¼
Buckwheat	9·	1·5	= 3·4	60·	2·	14·5	13·	1 : 7
Sunflower seed ...	13·	21·	=47·2	17·7	3·	37·3	8·	1 : 5
Linseed...	20·5	35·5	=79·9	15·2	3·4	13·3	12·4	1 : 4¾
Rice	7·6	0·3	= 0·7	77·4	0·7	...	14·	1 : 10¼
Ground oats	15·	5·5	=12·4	48·	2·5	19·	10·	1 : 4
Oatmeal	15·	6·	=13·5	62·	2·	5·	10·	1 : 5
Bran	14·5	4·	= 9·	51·5	6·	10·	14·	1 : 4¼
Pollard	15·	4·	= 9·	60·	4·5	3·5	14·	1 : 4¾
Barley meal	13·	2·	= 4·5	60·	2·	9·	14·	1 : 5
Bean and pea meal ...	25·5	1·5	= 3·4	45·	3·	11·	14·	1 : 2
Malt culms or sprouts	23·	1·8	= 4·	48·	6·8	10·	10·4	1 : 2¼
Brewers' grains ...	5·	0·4	= 0·9	9·4	1·2	7·6	76·4	1 : 2
Potatoes ...	2·2	0·0	= 0·0	20·3	0·9	0·6	75·	1 : 9
Clover hay	4·	0·8	= 1·8	11·4	2·	5·	76·8	1 : 3 3/10
Cabbage	2·4	0·4	= 0·9	3·8	1·4	1·5	90·5	1 : 2
Onions	1·5	0·2	= 0·5	4·8	0·5	2·	91·	1 : 3½
Lean meat	20·	3·	= 6·7	0·0	2·	0·0	75·	3 : 1
Green bone	20·	26·	=58·4	0·0	24·	0·0	30·	1 : 3
Milk	4·	3·5	= 7·9	5·	0·7	0·0	86·8	1 : 3½
Milk separated ...	3·9	0·4	= 0·9	4·5	0·0	0·0	91 2	1 : 1¼
Egg	14·	16·	=36·	0·0	1·1	0·0	68·9	1 : 2¼

The above has been compiled from several tables in order to arrive at a mean standard which would represent an average sample of fair quality.

The above references on the feeding subject from a scientific point of view are brief, but not unnecessarily so, seeing that were they given exhaustively to a degree they would be no more valued, for not one poultry man in a hundred would ever think of putting them into practice, and as mentioned.

before, were a diet formulated by either weight, measurement, or scientific value, other important essentials in poultry-breeding for profit might be overlooked, and the financial result disappointing. We all know maize is a very fattening food, its ratio being shown as 1 : 7⅓; yet the farmer, say, in the Wellington district who is breeding some fowls for the table, and has a quantity of small broken wheat, would never think of sending the latter to the market to be sacrificed for a few shillings a sackful, and pay market rates and railage on maize from Sydney. The latter would fatten his fowls quicker than the light wheat, but not in proportion to the cost.

In the same way the farmers on the Northern Rivers, who have maize in abundance, irrespective of its fattening tendencies, is not going to pay freight on it to Sydney and purchase wheat in return for his laying hens.

The whole aspect of the food question is in relation to the cost. Certain grains, roots, or meat may make the most balanced ration in the world, but the cost may be such as to make them prohibitive for feeding fowls. An illustration will suffice. A most excellent diet for fattening cockerels for the market is maize-meal and potatoes. This food is the most largely used in Ireland, where they do an immense export poultry trade to England. The former, although grown here, is not cheaper than in Ireland; but coming to potatoes, in this country of countless acres and every variety of soils, temperatures, and other favourable conditions, we do not produce enough for our own needs, with the result that this product is two or three times dearer than in the Old Country, and for that reason an unprofitable commodity to feed our fowls.

The following summary of the properties of the various foods will be found useful, their use by the poultry-keeper depending solely on the price on which they can be locally obtained.

The requirements of fowls, as omnivorous birds, are—

Grain, ground or whole; herbage, cooked or raw; flesh in some shape or other; water; grit and lime.

Grain is given either as hard or soft food. The latter, being a departure from nature, needs some explanation. Feeding on hard grain under domestication must not be compared with what a fowl does in this direction in a wild state, when the food is picked up here and there. In such case, by the time hunger is appeased a great portion of the first-found seeds have been softened and passed on from the crop to the gizzard, and have reached the consistency of soft food, whereas under domestication the fowl usually receives its food at stated intervals and in bulk. The crop is more quickly filled. So that the grain may be more rapidly assimilated, it is expedient to give soft food, especially at certain seasons of the year, as a morning meal.

Soft food, mixed dry and crumbly, should always be given in the morning in winter, as the fowls are thereby better enabled to withstand the difference between the outside atmosphere and that of the sleeping-house. It is well to give the soft food warm in cold weather.

Hard grain should form the evening meal, because, being more slowly digested, it carries the fowls better through the long hours of night.

Herbage plays such an important part in a fowl's dietary that to neglect the supply of it is most culpable. Where there are grass runs, greenstuff is at hand, especially in summer; but fowls in confinement often suffer for the want of it. Lettuce, cabbage, mangolds, turnips, either cooked or raw, will be beneficial and relished.

Chaffed clover hay, scalded overnight and mixed with the soft food, is excellent for laying hens, and is capable of supplying part of the lime needed for egg-shells.

Flesh is secured by fowls at large in the shape of worms, insects, and grubs, but in confinement and during winter this is denied, and meat should be supplied in some form.

Water must be given regularly and often. The vessels containing it must be frequently cleaned, and should be kept out of the sun's rays. A great many ailments contracted by fowls are caused by polluted water.

Grit is an absolute necessity. When taken into the gizzard the powerful muscles of this organ use it as mill stones, consequently it should be hard and rough.

Lime must be provided, and may most conveniently be in the shape of old mortar or broken oyster shells. This is needed, as before mentioned, for bone formation and for the shell of the egg. Egg-shells, if well broken, may be mixed with the soft food, or put in the lime-box.

Oats have the best balanced ratio of all grains, but for poultry they must be white and heavy.

Wheat is a good food, but deficient in fats and excessive in starchy matter. Bran and pollard are well-balanced foods, and most useful in poultry-feeding.

Barley is low in flesh and fat formers and excessive in fibrin.

Maize has been subject to much abuse as poultry food. It is somewhat low in flesh formers and high in fats (often more so than given in the foregoing table): it has also a high percentage of starchy matter, but it is extremely useful in balancing other foods and as an occasional food in winter for fowls at large.

CHAPTER XLII.

Conclusion.

HAVING now shown the many advantages which farmers possess enabling them to do well with fowls, there are other considerations which should still further assist in making this issue of the farm a very profitable one. In profitable poultry breeding, eggs are the chief stand-by. Any well-cared for flock of fowls of a good laying strain, young and healthy, will produce eggs all the year round; but no matter how well managed, nor when the birds be hatched, the larger number of eggs will be laid in the spring months, when they are cheapest. The laying competitions in this and other States have shown that individual hens have laid as many as 20 dozen eggs each, still a flock of the very best selected will fall far short of that number, there being undetected drones in every poultry-yard; consequently, it will not be fair to draw deductions from any individual fowl, nor from a single annual competition, but to take a series of such.

The Hawkesbury College have completed four annual tests, the average laying per hen for the respective years being 130, 160, 154, and 156, the average of these four tests thus showing 153 eggs, or 12¾ dozen for each hen, and the farmer or other poultry-keeper who secures 12 or even 11 dozen from each member of his flock during the twelve months will have a profitable return from his fowls.

Farmers' Fowls. The above six Black Orpingtons laid 1,461 eggs in twelve months.

The time when these eggs are produced is another important matter, seeing that the price for one or two months is as low as 7d. per dozen, and at other periods as high as 1s. 8d.; but, taking the year right through, the average of late has been about 1s. per dozen, and considerably higher than in any other State, the Queensland and South Australian markets only averaging 8d. or 9d. per dozen. Other markets are still lower, particularly the United States, where they can profitably produce eggs at 5 to 7 cents per dozen less than here. However, the farmer or other poultry-keeper in New South Wales has no need to accept 7d. per dozen for his eggs, for, unlike other countries, there is a Government here ever anxious to help the producers, which, in the matter of eggs, is in the provision made for cold storage. This system was incepted by the Agricultural Department in 1897. Its popularity and advantages can be best realised by the fact that during the past year 288,648 dozen were placed in cold storage, and at the end of five or six months profitably marketed, from 1s. to as high as 1s. 6d. per dozen being obtained.

As showing the extent to which farmers and others avail themselves
of this system of increasing their profits from poultry breeding, the
following figures, taken from the Superintendent's report, will be of
interest :—

1898	...	11,000 doz.	1901	...	140,292 doz.	1904-5...	251,640 doz.
1899	...	93,000 ,,	1902-3...		130,524 ,,	1905-6...	288,648 ,,
1900	...	96,000 ,,	1903-4...		151,128 ,,		

The storage of eggs fluctuates very much in ratio with the prices. In
January and February, 18 and 45 cases were received, while in March,
April, May, June, July, and August, none were put into store, the season
being practically closed; but, opening again on 1st September, 5,097
cases were received, followed by 2,654 cases in October, 166 in November,

Farmers' Fowls. The above six White Leghorns laid 1,443 eggs in twelve months.

and 38 in December, or a total of 8,018 cases. The deliveries, however,
were—in January, 261; February, 448; March, 1,208; April, 2,025; May,
2,285; June, 538; July, 15; August, 12; September, 10; October, 5;
November, 68; December, 403; or a total of 7,278 cases. The receipts
and deliveries, as shown above, afford a correct monthly index to the
market value of the product of the hen, one season being almost a
duplicate of its predecessor.

The spring months of September and October are those wherein all
poultry produce the greatest quantity of eggs, and, the markets then
being at their lowest, farmers and others largely confine themselves to
these months for storing; and, although with but the one object of holding
over till a dearer period, this has the additional effect of relieving the
overstocked markets in the months mentioned. Indeed, had cold storage
not been available during the past season, and the above 288,000 dozen
left on the local market, the result would have been disastrous to the
producers. Again, just as certain spring months in the year are the
cheapest, and those wherein the greatest storage is done, in the same way
two or three of the early winter months, particulary April and May, are

the dearest for this product, and those wherein the largest deliveries take place. At the same time, once February arrives, a distinct rise takes place, and from this on deliveries are made in increasing numbers till the months mentioned, June generally witnessing a clearance. This gradual and lengthened delivery, as opposed to the brief season of receiving, has the wholesome effect of the market never being overstocked with the Cold Store eggs to an extent of affecting the price of the current arrivals during the dear period of the year.

Farmers' Fowls. The above six Silver Wyandottes laid 1,303 eggs in twelve months.

The procedure in relation to the cold-storage system is simplicity itself. The producer in any part of the State has only to take his case or cases of eggs to the nearest railway station, and address them to the Government Cold Stores, Sydney, and pay freight; the Government collect the eggs and place them in cold storage, and when the dear winter months arrive deliver them to the depositor's order, to be disposed of where and how he desires, the charges of 1d. per dozen for twelve weeks to be paid on delivery.

Many country people wonder why there should be so many prices of eggs in the Sydney markets, and experience some difficulty with the various egg quotations of the daily papers. The following extract, from an article I contributed to the "Guide for Immigrants," will be explanatory on this subject :—" What are known as ' New Laids ' are from the suburban poultry-farmers, orchardists, and others who bring in a few dozen each sale-day. These are rarely more than one week old. They embrace small lots of from 3 to 10 or 20 dozen, and are usually purchased by boarding-house keepers, and grocers who have a special trade for the best goods. They bring 1d. to 1½d. per dozen more than any others offered. ' Railways,' as the name indicates, come by rail, mostly from farmers along the railway route. These are usually of good quality, and come next to new laids in price. Following these are 'Souths,' or South Coast. These come from the various ports of call south of Sydney, beginning at Wollongong and extending to Eden. 'Norths' or 'Northern Rivers' come not only from the North Coast, but in some instances 100 miles or more away up the extreme reaches of northern rivers—The Tweed, Clarence, Richmond, and others. In the winter season, not more than 2d. per dozen separates these from the suburban article, but in the hot weather they range as much as 4d. below. The 'Cold-room' eggs are those which have been in the chilling-rooms, say from September to the

Farmers' Fowls. Black Orpingtons. General utility; weight, 6 hens, 38½ lb. The six hens laid 1,188 eggs.

dear time in May and June, and usually bring from 1s. to 1s. 4d. or more per dozen. The following are the average monthly prices for the three principal lines during the past year :—

	Norths.		Souths.		New-laids.			Norths.		Souths.		New-laids.	
	s.	d.	s.	d.	s.	d.		s.	d.	s.	d.	s.	d.
January ...	0	11¼	1	0¾	1	3	July ...	0	11	0	11¾	1	2
February ...	1	0	1	1½	1	4¼	August ...	0	8	0	8¾	0	9
March ...	1	2½	1	3½	1	6	September ...	0	7¼	0	7	0	8¼
April ...	1	4½	1	6½	1	10	October ...	0	7	0	6¼	0	7¼
May ...	1	5¼	1	7	1	9½	November ...	0	5¾	0	7¾	0	9
June ...	1	2½	1	3½	1	6¼	December ...	0	7¼	0	8¾	0	10½

The average price for the past four years was :—In 1901, 1s.; 1902, 1s. 4d.; 1903, 1s.; 1904, 1s.; 1905, 11¾d.

The next advantage which the poultry-farmer here has over his compeers in other countries is the excellent all-the-year-round market for his poultry—indeed, for eight or nine months of the year the Sydney prices are considerably higher than what obtains for the same class of goods in England. Appended are the figures supplied me a few months ago by the four city poultry salesman, all showing the profitable nature of producing good table fowls for the Sydney markets :—

Haymarket, 5 July, 1906.

Sir,—Replying to your favour of the 27th ultimo, I have pleasure in forwarding herewith list of the top prices obtained by me between May, 1905, and May, 1906. Although these are the top prices realised in the months stated, I think it is only fair to point out that these figures were much more frequent during the months of August to March inclusive; while for the months of June, July, April, and May, they were exceptional prices, and not so frequent as in the other months. Further than this, I feel that it should not be lost sight of, these birds were sent in in the ordinary course of business, many of them coming to me by steamer, thus not having the benefit of the special conditions that birds prepared for a prize test for export would undergo. I frequently recommended my consignors to push on the growth of their cockerels, and market them at four to five months' old, at most, instead of holding them for two or three months longer for the slight advance in the price. I am fully impressed that our market demand is not so much for a large bird as for a medium-size plump bird of a tender age. You are at liberty to make use of my name in your report should you think fit.

Yours, &c.,
C. J. TURNER,
The Poultry Farmers' Exchange.

LIST of Prices, as advised per letter of even date.

1905.	s.	d.	s.	d.	s.	d.	s.	d	1906.	s.	d.	s.	d.	s.	d.	s.	d.
June ...	5	6	5	3	5	0	4	11	Jan. ...	6	9	6	6	6	0	5	8
July ...	6	3	5	6	5	4	5	3	Feb. ...	6	9	6	6	6	5	6	4
Aug. ...	6	6	6	1	6	0	5	9	March...	6	9	6	6	6	3	6	0
Sept. ...	8	0	6	7	6	6	6	0	April ...	6	11	6	10	6	7	6	4
Oct. ...	6	9	6	7	6	4	6	2	May ...	7	3	6	9	6	3	5	9
Nov. ...	7	0	6	11	6	9	6	7									
Dec. ...	7	9	7	4	7	0	6	11									

Sydney, 4 July, 1906.

Sir,—In reply to yours of the 27th instant, attached you will find information required. We might also inform you that we so rarely get sales of English ducks (Aylesbury, Pekin, &c.) that we could not report to you anything reliable. You have our permission to insert the firm's name in your article.

Yours, &c.,
ELLIS & CO.,
Poultry, &c., Salesmen.

HIGHEST Prices realised at our Sales.

1905.			s.	d.	1906.			s.	d.
May	...	Fowls (Roosters)	5	6	Jan.	...	Fowls (Roosters)	7	6
June	...	,,	5	10	Feb.	...	,,	6	11
July	...	,,	6	6	March	...	,,	6	5
Aug.	...	,,	6	8	April	...	,,	6	3
Sept.	...	,,	5	8	May	...	,,	7	4
Oct.	...	,,	7	6					
Nov.	...	,,	6	6					
Dec.	..	,,	7	6					

Sydney, 2 July, 1906.

Sir,—In accordance with your request, I have pleasure in forwarding a list of highest prices obtained for cockerels and English ducks at our weekly sales. We held no sales prior to October, and, consequently, cannot get behind that date. It should also be remembered that for the space of at least three months our prices were adversely affected by the buyers carrying on a boycott. I have put down the figures for each consecutive week.

Yours, &c.,
THOS. REID,
The Poultry Farmers' Co-operative Society (Limited).

EXTRACTED from Weekly Sale-sheets.

Date	Cockerels.					English Ducks.			
1905.	s. d.	s. d.	s. d.	s. d.		s. d.	s. d.	s. d.	s. d.
October ...	4 10	5 0	5 7	5 9	...	2 9	2 10	3 0	3 3
November ...	5 11	6 9	6 4	6 0	...	3 0	3 11	3 10	3 4
December ...	6 3	6 5	6 10	7 6	...	3 6	3 11	3 10	5 4
1906.									
January ...	6 2	4 9	6 0	5 9	...	3 11	4 5	3 9	3 7
February ...	6 9	5 7	6 5	...	...	3 1	2 8	3 4	...
March ...	5 3	6 1	6 0	5 10	...	3 1	2 10	2 10	3 0
April ...	5 3	5 6	5 5	5 8	...	3 7	3 7	4 0	3 10
May ...	6 0	5 9	6 1	5 6	...	4 4	3 9	5 6	4 3

Sydney, 9 July, 1906.

Sir,—In reply to yours, I have gone through my account sales journals with the following result, *re* best prices for fowls at auction from May, 1905, to May, 1906 :—

1905.	s. d.	s. d.	s. d.	s. d.	1906.	s. d.	s. d.	s. d.
May ...	3 10	4 4	4 1	4 6	Jan. ...	5 9	6 9	7 6
June ...	4 10	5 0	5 1	5 2	Feb. ...	4 9	5 3	5 9
July ...	5 0	5 2	5 3	5 8	March ...	5 0	5 6	6 3
August ...	5 4	5 6	5 9	6 0	April ...	5 6	6 3	6 7
September ...	5 10	5 11	6 2	...	May ...	5 9	6 3	6 6
October ...	5 11	6 0	6 9	...				
November ...	5 10	6 4	6 7	...				
December ...	5 9	6 6	7 9	...				

Yours, &c.,

W. F. MURPHY,

R. T. Murphy and Company.

Should prices such as above materially drop, there is always a market awaiting in England; and in this branch the Government again acknowledges its maternity to those on the soil, by preparing their birds for the oversea markets. These, like eggs, are to be forwarded, freight paid, to Sydney, where they are received by the officials there, killed, plucked, dressed, frozen, and placed on board at the inclusive nominal charge of $3\frac{1}{2}$d. per head for fowls and ducks. In the very plentiful and cheap times as many as 120,000 head of poultry have been exported; the high prices of later years, as shown above, having considerably contracted the export trade, which last year totalled but 7,955 fowls, 3,310 ducks, 850 geese, and 3,438 turkeys. However, should there be a continuance of good seasons, and the corollary of cheap foods, poultry may again increase beyond our own consumption, and the oversea market will again have to be resorted to; and as a guide to the prices to expect, the following is supplied by a leading poulterer of Leadenhall Market for the best quality of dead fatted fowls. The prices quoted are for each bird, and should our quality approach those quoted, there will be a deduction for freight, freezing, and other charges of from 9d. to 10d. per bird:—

Fowls.	January.	February.	March.	April.	May.	June.
Yorkshire	2/- to 3/6 each	2/- to 3/- each	2/3 to 3/6 each	2/3 to 2/6 each	2/3 to 2/6 each	2/6 to 3/3 each
Essex	2/- to 3/3 ,,	1/9 to 3/- ,,	2/3 to 3/6 ,,	2/3 to 2/6 ,,	2/3 to 2/6 ,,	2/3 to 3/6 ,,
Surrey	2/- to 4/6 ,,	2/3 to 5/- ,,	2/6 to 5/- ,,	2/6 to 5/- ,,	2/6 to 5/- ,,	4/- to 6/- ,,
Sussex	2/- to 4/6 ,,	2/- to 5/- ,,	2/6 to 5/- ,,	2/6 to 5/- ,,	2/6 to 5/- ,,	3/- to 6/- ,,
Welsh	1/6 to 2/6 ,,	1/6 to 2/6 ,,	2/3 to 3/- ,,	2/3 to 3/- ,,	2/6 to 3/- ,,	2/6 to 3/3 ,,
Irish	1/6 to 2/3 ,,	1/6 to 2/- ,,	2/3 to 2/9 ,,	2/3 to 3/- ,,	2/6 to 3/6 ,,	2/6 to 3/3 ,,
Hens (Old)	1/6 to 2/6 ,,	1/9 to 2/9 ,,	1/9 to 2/9 ,,	1/9 to 3/3 ,,	1/9 to 2/6 ,,	1/9 to 2/6 ,,
Capons	4/- to 5/6 ,,	4/- to 5/6 ,,	4/6 to 7/- ,,	4/6 to 7/- ,,	5/- to 7/- ,,	5/- to 7/6 ,,
Ducklings	2/6 to 4/- ,,	2/6 to 4/6 ,,	4/- to 7/- ,,	3/3 to 8/6 ,,	4/- to 8/6 ,,	4/- to 6/- ,,
Goslings	4/6 to 7/6 ,,	4/6 to 7/- ,,	5/- to 8/- ,,	5/6 to 8/6 ,,	5/- to 8/- ,,	5/- to 8/- ,,
Turkeys	8d. to 1/2 per lb.	8d. to 1/- per lb.	8d. to 1/- per lb.	8d. to 1/- per lb.	8d. to 1s. per lb.	8d. to 1/- per lb.
Eggs. English best new laid.	18/- to 20/- per 120	12/6 to 14/- per 120	8/- to 10/- per 120	7/- to 8/6 per 120	7/- to 8/- per 120	8/6 to 9/- per 120

Fowls.	July.	August.	September.	October.	November.	December.
Yorkshire	2/- to 3/- each	2/3 to 3/6 each	1/6 to 2/6 each	1/9 to 3/- each	2/- to 3/- each	1/9 to 3/- each
Essex	2/3 to 3/- ,,	2/- to 3/- ,,	1/9 to 2/6 ,,	1/6 to 2/9 ,,	2/- to 2/9 ,,	1/9 to 2/9 ,,
Surrey	2/6 to 5/- ,,	2/6 to 4/6 ,,	2/- to 3/- ,,	2/- to 4/- ,,	2/3 to 4/- ,,	2/3 to 4/- ,,
Sussex	2/6 to 5/- ,,	2/6 to 4/6 ,,	2/- to 3/- ,,	2/- to 4/- ,,	2/3 to 3/9 ,,	2/3 to 4/- ,,
Welsh	2/- to 3/- ,,	2/- to 3/- ,,	1/6 to 2/6 ,,	1/6 to 2/6 ,,	1/6 to 2/- ,,	1/6 to 2/6 ,,
Irish	1/6 to 2/6 ,,	1/9 to 2/6 ,,	1/3 to 2/- ,,	1/3 to 2/- ,,	1/3 to 2/- ,,	1/3 to 2/- ,,
Hens (Old)	2/- to 2/6 ,,	1/9 to 2/9 ,,	1/6 to 2/- ,,	1/6 to 2/6 ,,	1/6 to 2/6 ,,	1/6 to 2/6 ,,
Capons	5/- to 7/6 ,,	5/- to 7/6 ,,	4/- to 5/- ,,	4/- to 5/- ,,	5/- to 6/6 ,,	4/6 to 6/- ,,
Ducklings	....	3/- to 4/- ,,	1/6 to 2/6 ,,	2/- to 3/- ,,	2/3 to 3/3 ,,	2/3 to 3/6 ,,
Goslings	....	5/- to 7/- ,,	4/- to 5/6 ,,	4/6 to 7/6 ,,	4/6 to 6/- ,,	4/6 to 8/6 ,,
Turkey Poults	5/- to 8/6 ,,	8d. to 1/- per lb.	4/6 to 7/- ,,	4/6 to 8/- ,,	5/- to 12/- ,,	....
Turkeys	....	....	....	....	....	4/6 to 25/- ,,
Eggs. English best new laid.	9/- to 10/- per 120	10/- to 11/- per 120	10/- to 12/- per 120	10/- to 13/- per 120	16/- to 18/- per 120	17/- to 19/- per 120

Within the present year there has been a good deal of agitation in South Australia regarding the export of eggs to England, resulting in the Government financially assisting a trial shipment of 700 cases.

However, seeing that the egg imports into this State are annually increasing, amounting last year to the almost incredible number of 1,452,207 dozen, it is scarcely likely that for a very long time we will have any to spare for export. Should a time of over-supply arrive, the English market awaits us, and it should be known that these goods are disposed of there by what is known as the long hundred, of which the following table is explanatory :—

Per 120 or long hundred.

15/-	=	8	a shilling, or	1/6	per dozen.
13/4	=	9	,,	1/4	,,
12/-	=	10	,,	1/2¼	,,
10/-	=	12	,,	1/-	,,
9/3	=	13	,,	-/11	,,
8/7	=	14	,,	-/10¼	,,
8/-	=	15	,,	-/9¼	,,
7/6	=	16	,,	-/9	,,
6/8	=	18	,,	-/8	,,
6/-	=	20	,,	-/7¼	,,

From all the foregoing, it will be seen that we have an excellent paying all-the-year-round market for both poultry and eggs; but should a time in the distant future arrive when our local demand is more than met, then there is the English market to fall back upon. However, it will be an evil day for the poultry-man when he has to compete in England against the Russian peasants, who can payably produce four eggs for 1d., and make money by rearing 3 lb. weight chickens at 6d. each.

Our great distance and expense in reaching the English market is another handicap to the establishing of a permanent profitable oversea trade in these products, which, with other factors, prompts the hope that, in the interests of poultry-breeders, our local markets may continue in the future the highly payable ones they have done in the past.

Sydney : William Applegate Gullick, Government Printer.—1906.

FAVEROLLES HEN.

MODERN LANGSHANS.

LANGSHANS.

TYPICAL BLACK ORPINGTON.

TYPICAL BLACK ORPINGTON.

TYPICAL SILVER WYANDOTTES.

PENCILLED AND PARTRIDGE WYANDOTTE.

TYPICAL BUFF ORPINGTON.